**Knives, Swords, and Bayonets:
A World History of Edged Weapon Warfare**

by Martina Sprague

Copyright 2013 Martina Sprague

All rights reserved. No part of this book may be reproduced in any form or by any means, electronic or otherwise, without the prior written consent of the author.

Acknowledgements:

Front cover image pictures eight Gurkha men depicted in a British Indian painting from 1815. Image source: Daviddb, reproduced under Wikimedia Commons license.

Back cover image pictures Shoulder Flash of the 6th Gurkha Rifles, a regiment of the British Army. Image source: Purplemonkey, reproduced under Wikimedia Commons license.

Image source for horse logo (slightly adapted) on back cover: CoralieM Photographie, reproduced under Wikimedia Commons license.

TABLE OF CONTENTS

Some Notes about the Knives, Swords,
and Bayonets Series   4

Introduction   10

Historical Background   14

Kukri Origin and Handling   29

Gurkha Recruitment Practices and
Mercenary Status   45

"Bravest of the Brave," Born or Made?   56

Concluding Remarks   66

Notes   71

Bibliography   87

SOME NOTES ABOUT THE KNIVES, SWORDS, AND BAYONETS SERIES

Knives, Swords, and Bayonets: A World History of Edged Weapon Warfare is a series of books that examines the history of edged weapons in Europe, Asia, Africa, the Americas, and the Middle East and surrounding areas before gunpowder increased the distance between combatants. Edged weapons were developed to function in foot or mounted combat. The primary battlefield function often determined the specific design of the weapon. In poorer societies the general populace frequently modified agricultural tools into weapons of war. The techniques for employing these tools in civilian life translated into viable methods of combat. When the advent of firearms made certain edged weapons obsolete, close range combat continued to rely on foot soldiers carrying knives and bayonets as sidearms to modern artillery weapons. But even in ancient times edged weapons were seldom the primary arms, but were frequently employed as sidearms to long range projectiles. Rebel fighters of Third World countries have likewise used edged weapons extensively in near modern and modern wars.

The Knives, Swords, and Bayonets series of books takes a critical look at the relationship between the soldier, his weapon, and the social and political mores of the times. Each book examines the historical background and metallurgic science of the knife, sword, or bayonet respectively, and explores the handling characteristics and combat applications of each weapon. The author suggests that the reader make specific note of how battlefield need and geography influenced the design of the weapon, the type of warfare employed (guerrilla, rebellion, chivalry, pitched battle, skirmishes, mass armies, etc.), and the type of armor available to counter the blow of a knife or sword.

The historical treatment of edged weaponry could fill volumes. Because of the vastness of the subject, certain restrictive measures had to be applied in order to keep the series within a reasonable length while still giving adequate coverage. For example, the author has chosen to cover Chinese and Japanese but not Korean sword history. Every reader is thus bound to find some favorite details omitted. While many treatments of the subject focus exclusively on the technological aspects of weapons, this series also considers the political climate and the environmental or geographical factors under which the weapons evolved. Moreover, every culture, western or non-western, employs a number of subtleties that are exceedingly difficult to understand fully, unless one has spent time living in and studying the specific culture. The same can be said for every subculture (a culture within a culture), such as a military organization. The reader is reminded that, unlike science which is mathematically precise, history offers a broad range of perspectives on every issue.

The narrative the author has chosen to write portrays the development and dynamics of edged weaponry from ancient to modern times, including the soldier's training and his view of military service. The close relationship between military and political or social history also spurred the author's desire to examine the carry of edged weapons as symbols of military rank and social status. Rather than covering battles in their entirety, the author has elected to illustrate bits and pieces of particular battles that exemplify how the weapon in question was used. The book series comprises ten books arranged by weapon type, geographical area, or time period, and is designed to introduce the reader to the great assortment of edged weaponry that has been used with varied success in most regions of the world. Each book in the series is an entity in itself. In other words, it is not necessary to read the books in any particular order. Hopefully, the series will provide the reader with a solid foundation for continued study.

For her research, and in order to render an analysis that closely describes the dynamics of battle and the cultural aspects surrounding edged weaponry, the author has relied on a large number of primary and secondary source materials including historical treatises, artifacts located at museums, ancient artists' renditions of war in sculpture, paint, and poem, eyewitness accounts to the events in question, books, articles, documentaries, Internet resources, university lectures, personal correspondence, and direct hands-on practice with weapons in mock battles. Note that source material is often contradictory in nature. For example, swordsmen of the same era and geographical region frequently differed in their views with respect to the conduct of battle or the "best" type of sword or battlefield technique. The reader is encouraged to keep an open mind and consider the different possibilities that the soldier faced, and why he would emphasize a particular type of weapon or combat technique over another. The endnotes provide additional information, clarification, and exceptions to commonly propagated historical beliefs.

The author reminds the reader that despite their lethal features, edged weapons are not randomly chosen bars of steel that can cut and kill. The difference between victory and defeat often lies in the soldier's knowledge, skill, and fortitude; in how well he handles his weapon, but also in how well the weapon adheres to the laws of physics with respect to balance and motion. Studying metallurgic science is the key to understanding the relationship between the weapon smith and the soldier. The knife- or swordsmith thus carried part of the responsibility for the soldier's success or failure. Additionally, edged weapons were an integral part of the soldier's kit and often represented abstract qualities such as bravery and honor. By understanding the history of knives, swords, and bayonets, one will gain insight into the culture—the external and internal forces—that shaped the men who relied on these weapons in personal struggles of life and death.

8

Kukris and Gurkhas

Kukris and Gurkhas

Nepalese Kukri Combat Knives and the Men Who Wield Them

INTRODUCTION

The *kukri* (see also khukuri), pronounced *koo-ka-ree* with stress on the first syllable, is derived from Sanskrit and means razor. This curved and broad-bladed knife resembling a machete is used primarily in close range hand-to-hand combat. It has been the national weapon of Nepal and a part of the soldier's uniform since 1837 with the Gurkhas' struggle for control of the Kathmandu valley.[1] The Gurkha soldiers of the Nepalese army, well-known for their fierceness in combat, trained with the kukri until they could wield it as if it were an extension of their arm. The kukri in the hands of a skilled warrior has been termed nearly impossible to defend against.[2]

The Gurkhas were brought to the world's attention in the nineteenth century, when the territory ruled by Britain's Honourable East India Company in Nepal experienced a number of boundary disputes and Gurkha raids. Mutual respect developed between the Gurkhas and the British, who viewed the Gurkhas as disciplined and capable fighters training under the motto, "It is better to die than to be a coward."[3] Prior to India's independence from Britain in 1947, these "hill men" of Nepal could find honor only in the service as professional soldiers in the East India Company, or in the Indian army. While the Gurkhas would eventually become an integral part of the British army who regarded them as the ideal model of excellence, their own country viewed them as a spoiled and troublesome minority who "deserted their own officers in time of need in 1815."[4] This ill reputation stemmed in part from Britain's habit of employing defectors from the Gurkha army in irregular operations. Moreover, the fall of Kumaon in the Himalayas on the border to Nepal, and the agreement of surrender that the governor of Kumaon signed, were considered terrible misfortunes in Nepalese history.[5]

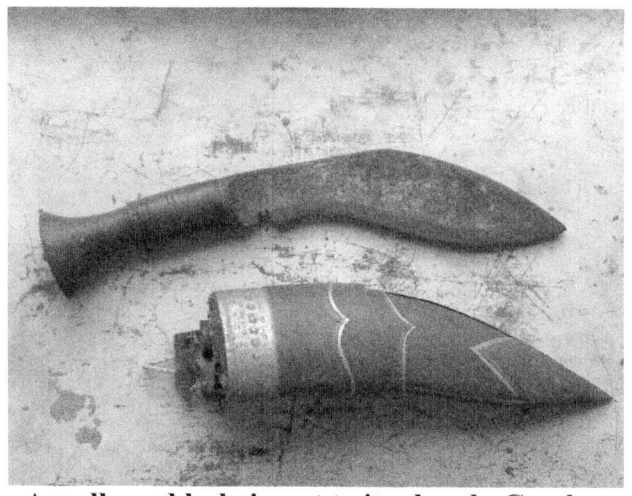

A well-used kukri next to its sheath. Combat knives exist in several shapes with different length blades. Exactly where the knife ends and the sword begins is debatable. A good fighting knife has a blade at least six inches long and often longer. Some knives, such as the Nepalese kukri, are long enough to be considered short swords. Image source: Bullygram, reproduced under Wikimedia Commons license.

In contemporary times the Gurkhas have served the British Crown in several campaigns worldwide, including both World Wars. As World War II ended Gurkhas would see active service mainly in the Far East, until Britain moved forward to defend the Falkland Islands following the Argentine occupation in the spring of 1982. As an important ally of Britain the Gurkhas made significant contributions to the campaign, albeit not in any larger battles. They have fought in every major expedition on foreign soil launched by the British since.

The Gurkhas have been termed the "only soldiers who can win their battles on their reputation alone."[6] Weapons shape the character of the men who wield them and two "upturned kukris meeting at the top, their naked blades crossing," serve as the identifying insignia of the Brigade of Gurkhas in the United Kingdom.[7] The western world has long had a fascination with this non-western fighting force. Do the Gurkhas deserve special attention for their achievements? Or are they merely minute cogs in a vast machine, no more special than the hundreds of thousands of soldiers who have passed through countless battlefields the world over? Retired Colonel John Philip Cross, who started his service with the Gurkhas in World War II and has lived permanently in Nepal since he left the army in 1982, suggests that the efforts of the Gurkhas often seem to go unseen and unheard.[8]

This book examines the kukri knife as a combat weapon and relates it to the history of the Gurkha soldier. It starts with a historical background of the Gurkhas, including an account of the campaigns in which they have fought. It then explores the origin and handling characteristics of the kukri and its combat applications. Next it discusses Gurkha recruitment into the British and Indian armies, followed by an examination of the status of Gurkhas as mercenary soldiers. The book summarizes with a discussion of the Gurkha reputation as the "bravest of the brave." The concluding remarks focus on the Gurkhas' so-called "inherent" fighting characteristics and the kukri knife as a symbol of status and skill.

HISTORICAL BACKGROUND

The Gurkhas[9] take their name from the eighth century (or thereabouts) Hindu warrior-saint Guru Gorakhnath. The term "Gorkha" comes from two Nepalese roots: "go" (cow) and "rakh" (protector). Because of their Hindu origins, the Gurkhas regard themselves as the "protectors of cows."[10] They have their roots in the Gorkhali army from the area of Gorkha, a Nepalese feudal village located on a hill overlooking the Himalayas approximately 144 kilometers (90 miles) northwest of the Kathmandu valley. In the eighteenth century the mountain kingdom of Gorkha grew in size when the kings and generals of the House of Gorkha conquered the nearby areas, and planned further expansion northward toward Tibet and southward toward India. The first Gurkha regiments were created in 1763.[11]

The western world's contact with the Gurkhas started when the East India Company sent an ill-fated expedition to assist the Newars, an indigenous people who inhabited the Kathmandu valley, in their struggle against the invading Gurkha armies in 1767. The Chinese and British were alarmed due to the aggressive Gorkhali expansion. Furthermore, trade agreements with the Newars and a possible alliance between Nepal and the Sikhs worried the British, who desired to expel the Nepalis from the region.[12] In 1768-1769 Kathmandu surrendered to the invading King of Gorkha, Prithwi Naraayan Shah, who proceeded to lay the foundations for the present state. The king "formed the Gorkhali Army, which helped him in succeeding to unite what is now Nepal into one kingdom."[13] The king's ability to defeat a much larger force is attributed in part to the unusual kukri knife, which after this event became the preferred weapon of the Nepalese troops; a tradition that has continued through the Gurkhas' service in the British and Indian armies.[14] Many people view the modern Gurkhas as descendents of those fighters who created the present Nepalese state. Gurkha soldiers have been mentioned in writings for at least 170 years, mostly through their contributions to British military history.[15]

From the mid-eighteenth to mid-nineteenth centuries, the Indian kingdoms and the British-led East India Company attempted to form a hybrid military organization that would better serve their purpose in the region. "The Gurkha Kingdom of Nepal copied the infantry system of the EIC [East India Company]. The infantry were drilled and organized in battalions. Deserters from the EIC's Bengal army taught the Gurkhas Western military discipline."[16] In the early nineteenth century during Britain's attempts to consolidate its powers in India, the expanding influence brought Britain into further contact with the Nepalese Gurkhas. The Gurkhas who were then the ruling authority in Kathmandu—at "the zenith of their power" they were "rubbing the edges" of the ill-defined boundaries between Nepalese territory and that conquered by the British—maintained their distance from the British to the extent possible. Relations between Nepal and the East India Company deteriorated.[17] Tensions mounted and raids followed along the borders, until the conflict culminated in war in 1812 between "the British East India Company and the Gurkha tribesmen."[18] In

November 1814, as a result of continued border tensions—the Gurkhas were marauding and seizing villages and attacking police posts, killing a number of Indian policemen—the British governor-general Warren Hastings, employing a force of 22,000 men, declared war on Nepal.[19] The first campaign of the British against the Gurkhas gave birth to the fierce reputation that has accompanied the Gurkhas into present day. The Gurkhas shunned the idea of retreat. A graphic example is how one Gurkha, badly wounded in the jaw by a musket ball, went to the British and asked for medical assistance. When he had been treated he refused to surrender, returned to the Gurkha forces, and continued to fight.[20]

Although the Gurkhas proved victorious in campaigns in Tibet and China, those serving in India against the forces of the East India Company in the Anglo-Nepalese War of 1814-1816 saw less success. Nepal lost approximately one-third of its territory to Britain. The Nepalese prime minister Bhimsen Thapa (reigned 1806-1837) increased the size of the Gurkha army and strengthened the military.[21] In an act of strategic necessity after reaching a stalemate in the war, the British governor-general decided to bring the Gurkhas under British command. Two powerful forces in the region would inevitably lead to confrontations, and the goal of making Nepal a peaceful neighbor of British India required the weakening of the Nepalese army. It thus seemed wise to employ the Gurkhas in the British-Indian army in order to drain soldiers away from Nepal.[22] A conflicting view is that the British conquest of India in the nineteenth century left Nepal, which was sandwiched between India and China, with little alternative but to befriend the British for the sake of preserving its independence and autonomy in domestic affairs. Regardless of which view one takes, in the end Nepal had to sue for peace. Impressed with the Gurkhas' fighting prowess, the British were granted the right to recruit Gurkha soldiers as mercenaries into the army of the East India Company. In return for this privilege, Britain guaranteed Nepal protection against foreign and domestic enemies.[23]

Balbhadra Kunwar, Nepali commander of the Anglo-Nepalese War of 1814-1816, which ended with British victory. Note the kukri in his left hand, with the concave cutting edge facing forward. Image source: Raymond Palmer, reproduced under Wikimedia Commons license.

The British considered the Gurkhas "the best soldiers in India with unadulterated military habits."[24] A British general wrote in 1815, "They are hardy, cheerful, and endure privations, and are very obedient, have not much of the distinction of caste and are a neutral kind of Hindu. Under our government, they would make excellent soldiers."[25] Through the Treaty of Sagauli signed in 1816 the Gurkhas became officially recognized for their military capabilities, mainly their courage, self-restraint, loyalty, and professionalism.[26] At the signing of the treaty Britain decided to create the first Gurkha regiment based on the Gurkha soldiers' bravery and fighting skill, and recruited large numbers of Gurkha volunteers for service in the East India Company. The first three thousand Gurkhas who served under British flag were divided into four battalions originally assigned to keep the peace in India.[27] Although focusing on the Far East, they were based in the United Kingdom. In the decades that followed, the hill villages of Nepal would provide thousands of Gurkha fighters for the British Empire.

The Gurkhas served in several wars in the nineteenth century, including the First and Second Sikh Wars of 1846 and 1848 where, "armed with the short weapon of their mountain, [they] were a terror to the Sikhs throughout the great combat."[28] During the Indian Mutiny of 1857-1859 and the rebellion of Indian troops against British rule in India, the mutineers offered "to pay ten rupees for a Gurkha's head," after which the British began taking even greater interest in the Gurkhas.[29] The British did no longer trust the Bengal army and started recruiting a variety of people from different provinces, in order that no particular group would have the power to unite in another uprising. Regimental officers exercised a balanced recruitment policy of "divide and rule," for the purpose of preventing groups from forming along religious lines. The Gurkhas from

Nepal was one of the groups recruited by the British; however, they continued to serve mainly in remote areas.[30] The 60th Rifles, later part of the Royal Green Jackets (an infantry regiment of the British army), "fought alongside the [Gurkhas] and were so impressed that following the mutiny, they insisted that the 2nd Gurkhas be awarded the honor of adopting their distinctive green uniforms [and] Rifle Regiments traditions, and that they should hold the title of Riflemen rather than Sepoys. From this point onward, the Gurkhas became an integral regiment of the British Army."[31]

The Nusseree Battalion, later known as the 1st Gurkha Rifles, circa 1857. Note the kukri on the belt of the soldier sitting on the left. Image source: British Army, reproduced under Wikimedia Commons license.

When Britain conquered and annexed all of Burma to British India in 1886 in response to French infiltration, large numbers of Gurkhas from the tribal villages were recruited to serve the British. The Gurkha tradition now began to be handed down through generations. One Gurkha soldier who served the British in World War II explains that from 1890 to 1915, both his grandfather and granduncle served in numerous expeditions against the rebellious tribes in Burma. The same soldier's father and uncle later fought rebellious tribes on the northern Burma frontier, in northeastern India, and in southwestern China in the early twentieth century.[32] Another Gurkha serving in the Falkland Islands campaign in 1982 tells how his father served in Burma in World War II, and how the Gurkha tradition of Regimental service extends through several generations.[33]

In World War I, 120,000 Gurkhas joined the army and were sent to the trenches "where 15,000 perished."[34] They fought in France, Flanders, Mesopotamia, Persia, Egypt, Palestine, and Salonika (Thessaloniki, Greece). They fought in the Middle East, most notably at Gallipoli—the British had proposed that a new front be opened at the Gallipoli peninsula south of Constantinople for the purpose of driving the Turkish enemy back and gaining access to Germany from the south, and also to the Black Sea[35]—where they suffered heavy casualties and had little reinforcements, yet furthered their reputation as fierce warriors by being "the only regiment to break through the Turkish lines," an event which became a defining moment in Turkish war history.[36] Gurkhas were also the first British units to break the German line at Neuve Chapelle in the spring of 1915.

According to one captain of a Gurkha Brigade, "At 7:30 a.m. artillery bombardment commenced, and never since history has there been such a one. You couldn't hear yourself speak for the noise. It was a continual rattle and roar. We lay very low in our trenches, as several of our guns were firing short."[37] In the second battle of Ypres, although met by heavy artillery fire, a few Gurkhas serving the British managed to fight their way through the enemy line using kukri and bayonet. In the end, however, the German soldiers got the best of them and killed the Gurkhas to the last man.[38] At the capture of Nasiriya on the Euphrates River they fared slightly better. A Gurkha attack was about to falter when a young Gurkha started swinging his kukri and dashed toward the Turkish trenches calling for his comrades to follow. They killed thirteen Turks with their kukris.[39]

In 1919 the Gurkhas fought in the Third Afghan War. They continued their tradition serving the British in World War II when Britain employed forty Gurkha battalions totaling approximately 112,000 men, fighting side-by-side with British and Commonwealth troops in Syria, the Western Desert, Italy, and Greece. At their height the Gurkha regiments stood at 250,000 men and fifty-five battalions.[40] These Nepalese soldiers carried the kukri and used it at opportune moments, killing "quickly and cleanly—like king cobras," whenever infantry combat took place at close range between individual soldiers.[41] The China-Burma-India Theater experienced overwhelmingly many hand-to-hand battles involving Gurkhas. *Gorkhali Ayo!* (The Gurkhas are coming!), or *Ayo Gorkhali*, chaaaaarge! (Here come the Gurkhas, charge!), if given as a command within the regiment, became the dreaded cry feared by the enemy troops.[42] During the battles on the Indo-Burmese border in 1944, the combined British, Indian, and Gurkha infantry forces supported by tactical airpower, "ground the Japanese 33rd Division and its supporting formations to pieces."[43] A Gurkha platoon from the 10th Regiment killed 125 Japanese defenders while suffering only two combat casualties of their own.[44] The overall death toll proved heavy, however. The Gurkha Brigade suffered 43,000 casualties in the two World Wars and won twenty-six Victoria Crosses. The Victoria Cross is Britain's highest award for courage in the presence of an enemy and is awarded for conspicuous bravery, resourcefulness under adverse conditions, and contempt for danger.[45]

In January 1948, as a result of the partition of India (known as the Tripartite Agreement) and "negotiations between the Nepalese, British, and Indian Governments, four Gurkha regiments became an integral part of the British Army, forming the Brigade of the Gurkhas."[46] In modern times the Gurkhas have demonstrated their jungle fighting abilities throughout the Malayan Emergency (1948 to 1960), using stealth tactics and claiming victories through attrition. India has used Gurkhas in the war against China in 1961-1962 and Pakistan in 1965 and 1971.[47] In British service they were called to action in Borneo in the Brunei Revolt in December 1962, "when rebels linked to the opposition People's Party attempted to seize power."[48] They saw continued action in a confrontation against Indonesian forces in Malaysian Borneo in 1963-1966, in the Falkland War of 1982, and in the Gulf Wars of 1991 and 2003.[49] In

2003 Britain employed 3,443 Gurkhas serving mainly in the infantry as an integral part of the British army while adhering to their Nepalese identity and culture.[50] When word came that he would be sent to Iraq, a Gurkha soldier who had served Britain for three years admitted that it could be a fierce war.[51]

British Gurkha soldiers in Iraq, 2004. Gurkha military members Mr. Tej Bahadur (left), Mr. Chitra Bahadur Chhetri (right), and Mr. Gopal Gurung on the tower, patrol the compounds of the Coalition Provision Authority (CPA) in Baghdad, Iraq. Image source: SSGT Quinton Russ, United States Air Force, reproduced under Wikimedia Commons license.

Gurkhas were fighting in Afghanistan as recently as November 2007, as part of a joint operation between Gurkhas, Canadian soldiers, and the Afghan National Army in an effort to "eliminate the source of rocket fire that targeted two Kandahar military outposts."[52] They have served in several other capacities including policing activities; for example, involving illegal immigrants in Hong Kong where one Gurkha soldier, armed with baton and kukri, reportedly countered and arrested a dagger-wielding illegal immigrant intent on avoiding capture.[53]

The kukri gave (and gives) the Gurkhas mental superiority over their enemy. Although Gurkhas serving the British were (and are) equipped with modern weapons including rifles, machineguns, and grenades, they have caused mobs to dissipate without bloodshed just by unsheathing their knives. Historically they have used their kukris as melee weapons to sever limbs and decapitate the enemy. A Gurkha officer explained that when the Gurkhas go into battle they cannot be stopped. The Argentine troops, at the sight of the kukris the Gurkhas carried as sidearms in the Falkland Islands campaign, are said to have "melted away and refused contact."[54] The worst fear among the enemy was not the actual battle *per se*, nor the size of the opposing army or their firing weapons and grenades. Rather, it was the Gurkhas' fighting ability which was described as second to none, their reputation for fierceness, and accounts of their use of the kukri for decapitation purposes that caused the enemy to surrender.

Wild rumors circulate around the Gurkhas. According to the memory of the Argentine survivors, the Gurkhas "preceded the English in the battle" and "advanced shouting and cutting throats" at a speed of one decapitation "every seven seconds." Blinded by rage the Gurkhas were unable to distinguish between friend and foe. They "continued killing even the Englishmen until these had to resort to the ultimate to overcome them." Although this account, based on personal memory of the event, has no doubt been exaggerated, it can nevertheless be established that the Gurkhas had a great psychological impact on the Argentine army.[55] An Argentine chaplain reported that "Gurkhas had slit the throats of some forty Argentine soldiers."[56] Another Argentine troop related how they were convinced that the kukri was not only a close-range decapitation weapon used in trench warfare, but that every Gurkha also had the skill to hurl his kukri and hit the target with accuracy from a distance of 100 meters (328 feet).[57] To further hone their mental edge, the Gurkhas displayed an air of confidence about how the battalion would fare against the Argentines. Prior to embarking on the campaign they could be observed "sharpening their kukris on a large millstone outside one company armoury." One Gurkha soldier exclaimed that there would be no problems; the enemy would be chopped "like radishes cut into little pieces."[58]

Despite these attempts to unnerve the opposition, all Argentine soldiers did not allow the mental game to victimize them. While some maintained that the Gurkhas had appeared disorganized and advanced as if "doped, stepping on Argentine mines, yelling like madmen,"[59] the psychological warfare, the picture painted of the Gurkhas as monsters who could not wait to behead their enemies with the dreaded kukri, according to one Argentine rear admiral, was the sort of history that cowards invent.[60] *The Daily Telegraph*, a renowned newspaper of the United Kingdom, reported that the Argentines attempted to project their own image of fierceness, explaining that their "eighteen-inch gaucho's knife" would easily annihilate the Gurkhas and their "foot-long daggers."[61] (Gauchos were South American horsemen or herders that can be viewed as the equivalents of American cowboys.) By contrast, one British soldier serving with the Gurkhas in the Falklands stated that he was unable to detect any gaucho knives among eighty-three prisoners of war who were forced to march in file, "disheveled and exhausted," for eight kilometers back to the headquarters.[62]

When attempting to determine the extent to which the knife affected the mental composure of the enemy, one might consider that the Argentines suffered from low morale much as a result of being poorly equipped. The lack of food robbed them of their strength, particularly in the cold weather which complicated the process of cooking what little food they had. The better equipped Gurkhas took advantage of the situation and threatened the depleted Argentine prisoners, telling them that any escape attempt would be stopped through the active use of the kukri. Although the threat was delivered in Gurkhali, a language which the Argentines could not understand, the sight of the sharpened blade and the body language of the (only) five-foot two-inch Gurkha who uttered the threat were enough to drive home the point.[63] Although the kukri was carried in every assault, it was likely withdrawn from its scabbard for the purpose of killing an enemy less often. A safe but not too distant observer might have noticed instead a plethora of modern weapons ranging from rifles and machine guns to grenade launchers. Moreover, the kukri, like other short range weapons such as the bayonet, proved nearly useless against long distance firing.[64]

Despite the fact that the Gurkhas are equipped with modern weapons and train and fight alongside western forces, without the kukri they are incomplete. When they run out of bullets or when their weapons malfunction, the opportunity remains to unsheathe their knives and continue fighting in hand-to-hand combat. A graphic example is a recent training exercise in the desert of the sultanate of Oman, where Gurkhas were observed to draw their curved razor-sharp kukris and enter hand-to-hand combat while simulating slitting their "opponent's" throat.[65] In a Gulf War battle preparation exercise, the American forces had difficulties distinguishing Gurkhas from Egyptians because of their physical similarities: their short, stocky builds and their skin color. When the Gurkhas whipped out their kukris, they removed all doubt as to their identity.[66]

The Gurkhas grew up carrying the kukri as a tool and weapon in their everyday lives. They learned the subtleties of the knife and came to identify with it to the extent where man and weapon became one. In order to understand the accomplishments of the Gurkhas and their impact on military history, the kukri must therefore be examined along with the soldier.

Indian army soldiers with 1st Battalion, 1 Gorkha Rifles take up security positions outside a simulated combat town during training exercises at the National Training Center at Fort Irwin, California, March 12, 2008, as part of Exercise Shatrujeet 2008. The combined arms exercise conducted by U.S. forces and members of the Indian army was designed to share knowledge and build interoperability for possible future operations. Note the modern firearms that are part of the equipment. Image source: LCPL Kevin McCall, reproduced under Wikimedia Commons license.

KUKRI ORIGIN AND HANDLING

The infamous kukri knife (also called a short sword) has a long history as a weapon and cultural symbol of the Gurkha soldier. The oldest known kukri dating to 1627 belonged to Raja Drabya Shah, King of Gorkha. But kukris likely existed prior to this date and are believed to be at least five hundred years old. Several weapons depicted in sculptures from as early as the third century CE closely resemble the modern kukri.[67] The knife was possibly introduced to India and Nepal even earlier through the ancient Greeks "by Alexander's Macedonian army, which invaded north-west India in the fourth century [BCE]."[68] The kukri might have descended from the classic Greek sword or knife known as *kopis* about two thousand five hundred years ago, which was carried in the armies of Alexander the Great. The peculiar broad back, wedge section, and concave cutting edge of the kopis strikingly resemble the kukri, which, when held vertically with the cutting edge facing forward, has been likened to a cobra ready to attack.[69] However, "it is also possible that the kukri is simply a design native to the hills of the Himalayas," where the knife has long served as an agricultural tool used for chopping wood, cutting dense forest jungle, and skinning animals.[70]

The kukri's sturdy design and the way it is weighted makes it a deadly close combat weapon as well as a durable and powerful field knife that can function as a utility tool in a combat environment, much like a machete, hammer, or shovel.[71] Gurkhas are known to have cut down a tree, lopped off the branches with the kukri, and used the remaining log as a battering ram. An example of indirect combat use of the knife involves a Gurkha being mauled by a leopard and escaping by striking the animal with the knife still in the scabbard.[72] During the Falkland Islands campaign, the Gurkhas used the kukri along with shovel and pickaxe for digging trenches in a hurry. They also used it for such tasks as slashing open tubes of camouflage cream.[73]

Fighting with edged weapons has proven exceedingly dangerous. Although knives have on occasion been used as bayonets to extend the soldier's reach, a knife is designed primarily for very close range combat between individual soldiers.[74] The versatility of the kukri, and the relative ease with which it can be handled even by an untrained person, is attributed in part to its simple design: A knife that lacks moveable parts is not apt to fail. In World War II Burma, for example, the Gurkhas came into conflict with the Japanese. When an enemy machinegun jammed, a Gurkha soldier attempting to fire back also experienced a jammed weapon. Resorting to his kukri, he managed to kill the Japanese soldier. To further demonstrate the lethality of the knife, in the same theater of war when the Japanese attacked from all sides with small arms fire, a "free for all" developed involving kukris, bayonets, and swords. The confusion is said to have been so severe that both Gurkhas and Japanese were killing their own comrades by mistake.[75] When a Gurkha pulled the body of a comrade from a bunker, he discovered seven dead Japanese soldiers who had been killed with the kukri before the Gurkha had been bayoneted to death.[76]

The heavy-duty blade of the kukri is designed primarily for chopping. A distinguishing characteristic of a chopping knife is the curved edge which is always presented at an angle to the target, regardless of which part of the blade is used for impact. Blade length varies but is typically between ten and fifteen inches. Some blades are as long as thirty inches; however, knives designed specifically for combat are generally shorter and lighter in order to preserve their effectiveness for close range fighting in tight spaces. The design of the kukri increases the cutting momentum of the blade. Kukris with particularly wide tips shift the weight away from the handle, adding force to the cut and permitting heavy blows with little effort. In contrast to lighter knives designed for maneuverability, the kukri is designed to end the fight with a single slash or chop.[77] The heavy and blunt outer edge balances the curved inner edge and makes the knife an ideal decapitation weapon. According to one Gurkha sergeant, the shape of the kukri symbolizes the three Hindu gods of Brahma (the Creator), Vishnu (the Caretaker of Heaven), and Shiva (the Destroyer of Evil).[78]

Although the strength of the knife ultimately lies in the hand and skill of the wielder, once the knife is set in motion little energy is needed to wield it in the proper direction. In short, the knife "acts" as though it "wants to cut." A drawback is that the momentum of the broad-bladed tip makes it difficult to recover the knife when swinging it with full intent. If the soldier misses the target, he risks cutting his own body or free hand through the sheer momentum of the swing. Care must therefore be exercised to keep the free hand away from the knife's trajectory.

(A soldier can use his free hand to reinforce the weapon hand, to wield a second weapon, or as a defensive check against his opponent's attack.) Because of these design characteristics, if the kukri is mishandled it can easily injure the soldier. Even a dull blade can potentially cut the flesh to the bone. Accidental self-inflicted wounds such as a cut from a dropped knife were not uncommon in combat. The tip-heavy design gives the weapon a tendency to fall straight down if dropped. The tip can easily penetrate the leather in a soldier's boot.[79]

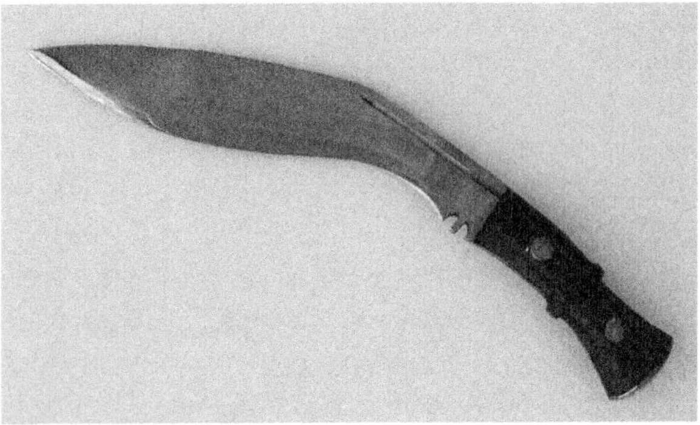

The kukri, often likened to a king cobra ready to attack, has been the national weapon of Nepal since 1837. The thick spine and broad tip make the kukri an excellent chopping weapon. Note the fuller stretching from the handle along the straight edge of the spine, and the notch on the blade near the hilt (called a *cho*) used for catching an opponent's weapon. Fullers reduce the weight of the blade without reducing its strength. The fuller is less crucial on knives than on swords of greater length and weight. Sometimes fullers on knives are purely decorative in purpose. Image source: Martina Sprague.

Most kukris were (and are) made in Nepal or India, traditionally by skilled craftsmen known as *Kami* (an untouchable caste). The superb craftsmanship has ensured that a kukri has never been broken in battle.[80] Since World War II, mass production of kukris has become more common. In order for the knife to meet its function as a hacking weapon, the steel must be of superior quality and able to hold a sharp edge. Although scrap metal was used during shortages in World War II to produce quite satisfactory weapons, the material of choice is carbon steel, normally a chunk of railway track because of its quality and strength. A slice from the rail is repeatedly heated and hammered on an anvil into its proper shape. When the final shape is acquired, the metal is heated again. The blade is then fine tempered and made stronger through a controlled cooling process, traditionally "with water poured from a teapot."[81] Different craftsmen are employed in the construction of the handle which is normally made of wood, and sometimes of buffalo horn or metal. The blade is secured through the tang and attached with rivets through the hilt. (A tang is an extension of the blade inserted through the handle.) *Laha*, a Nepali gum, is used as a sealant. The blade and point are then sharpened and polished.[82]

The semicircular notch in the blade near the hilt, called a *cho* or *kauro*, increases the weapon's defensive capabilities by allowing the wielder to block and catch an enemy's sword blow or bayonet thrust. By tilting and turning the cutting edge, the soldier can deflect or diminish the force in his opponent's strike instead of meeting the blow head-on. Or he can catch an enemy's saber or dagger in the notch and disarm his opponent with a quick twist of the knife.[83] The notch thus has a combat application similar to the *quillons* that were popular on the cross guards of medieval western swords. The notch is also said to interrupt the flow of blood onto the handle, so that the soldier can retain a good grip after he has used the knife against an enemy. Philosophically the notch represents the sun and moon, the symbols of Nepal, or the female generative organ, which is said to give the blade strength and efficiency.[84] "The Gurkhas believed that a kukri used to kill an enemy captured a part of his soul, thus becoming a spirited sword."[85]

The kukri has three primary uses: disarm, disable, and kill. Although the design lends itself to chopping, a trained wielder can use several strikes. While the primary cuts with most combat knives comprise the chop, slash, and thrust, the kukri has three secondary strikes as well: the hammer with the back (spine) of the blade, the slap with the flat of the blade, and the butt with the pommel or bottom of the handle.[86] These strikes utilize a part of the knife that is dull and are therefore considered blows rather than cuts. Blunt blows with the side of the blade or the butt of the weapon are generally used to disarm an enemy. Sharper blows to the head with the spine of the blade or slashes to the body are used to disable an enemy. Although many single edged knives have a "false edge" (a part of the spine that can be sharpened as a second edge), the kukri hardly ever has this feature. Its exceptionally thick spine would prevent it from being sharpened. The spine can thus be used effectively as a striking weapon. The tang which normally extends through the handle adds additional stability to the knife.[87]

In order to kill, the soldier applies a full force swing with the cutting edge directed at the target. Most killing or incapacitating blows comprise only one of three moves: a head strike, a gut strike, or a leg strike. After the initial cut has been made with the part of the blade closest to the handle, the curved tip allows the soldier to draw the weapon back toward himself while slicing through the target.[88] This type of combined chopping and slashing motion can be used, for example, to decapitate an enemy or cut off a limb. The Gurkhas have a reputation for slicing a "human head off with a single knife stroke."[89] A Gurkha soldier fighting in World War II Burma claims to have hidden behind a rock while his comrades were driving the Japanese enemy away. As the Japanese came running past the rock, he cut the heads off them one by one, decapitating eighty Japanese soldiers with the kukri in a single battle.[90] Although the kukri has been described as having its greatest combat use as a decapitation weapon, the soldier would naturally also target his enemy's torso and limbs.[91] It might be noted that the vivid decapitation stories do not necessarily make the Gurkhas more bloodthirsty than their enemies in battle. Due to their extensive use of the kukri as a utility tool, the Gurkhas were simply more comfortable with the idea of drawing the knife than with fixing a bayonet to a rifle.[92]

The Gurkha carries his kukri in a sheath attached vertically to his belt, normally on his left side if he is right-handed with the concave cutting edge facing forward. The knife can also be carried vertically at the belt along the soldier's back or diagonally through the belt at the belly with the cutting edge facing downward. Carrying the knife upright along the back gives the soldier unrestricted movement of his arms when the knife is sheathed. This carry position makes a quick draw more difficult, however, and is therefore normally reserved for ceremonial purposes.[93] The kukri sheath is made of wood with a leather covering. Unlike bayonets which can rust in damp climates if sharpened and lose their protective coating against corrosion, the kukri generally does not have a corrosion problem, because of its wood and leather scabbard and the time the soldier spends to hone and sharpen the weapon.[94] Along with the kukri the soldier carries two small knives, called the *karda* (a sharp utility tool also used as an eating utensil or to cut meat from bone) and *chakmak* (a dull knife used for sharpening the kukri or for making sparks for lightening a fire).[95]

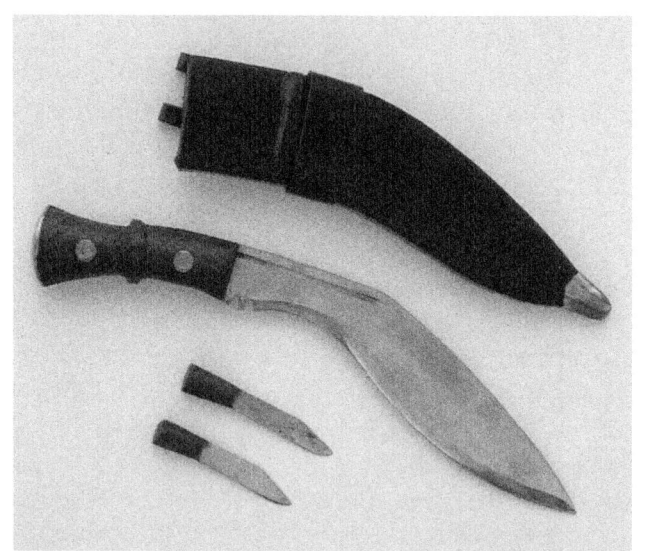

The kukri pictured here has a 12-inch blade and a 17-inch overall length. Note the protective metal tip on the sheath and the two smaller knives, the *chakmak* (a dull knife used for sharpening the kukri) and *karda* (a sharp utility tool). Image source: Martina Sprague.

Normally, the longer the knife the more difficult it is to wield successfully. A long weapon such as a sword, for example, limited by its own inertia, cannot change direction or be recovered as quickly as a short weapon; it is difficult to draw and wield with one hand and requires a great deal of training to use successfully on the field of battle. By contrast, a drawback of a relatively short weapon such as the kukri is that the soldier must be exceptionally close to his enemy, often within an arm's reach in order to accomplish his goal. The closer the range the more personal combat tends to become and, as is generally accepted, a greater degree of courage is required to engage and kill the enemy. Wielding the knife without proper spirit could have a negative psychological effect on the soldier and prevent him from doing the damage that his combat situation calls for.

Since it takes only one good chop to kill an enemy, the Gurkha must seize the initiative and attempt to land the first strike. Speed in drawing the weapon is crucial. The soldier utilizes a cross-draw when unsheathing the knife; he grabs the back of the sheath with his free hand and tilts it slightly forward. (Cross-draws are normally used with longer weapons such as large knives or swords, in order to facilitate ease of draw. This is also why the soldier wears his weapon on the left side of his body if he is right-handed, and vice versa.) Due to the curved design of the knife, he leaves the inner surface of the scabbard in contact with the spine to provide guidance throughout the draw. If the soldier is not careful, a maneuver as simple as drawing the knife can have devastating consequences. If he fails to use his free hand to guide the draw or wraps his fingers completely around the scabbard, the knife can cut through the leather and into the soldier's hand. The hand should therefore be kept away from the edge and toward the spine during the unsheathing process. Although the Gurkhas have a reputation for drawing blood (killing an enemy) every time they remove the knife from the scabbard, this "myth" has only held partly true when the knife has been used in sacred ceremonies, as "when the Gurkhas nicked their fingers and smeared blood on their swords."[96]

During training exercises and combat encounters, the soldier holds the kukri in a tight closed-fist grip with the fingers and thumb completely enveloping the handle, the thumb resting on top of the index finger, and the cutting edge facing forward or downward. A variation is to extend the thumb along the spine for greater dexterity. This grip is used, for example, when skinning animals, but is normally not used in close range combat that requires a secure grip for wielding the knife with power.[97] Moreover, for defensive reasons it is generally not a good idea to extend a finger or thumb along the blade while using the knife in battle, where one's hand can become an easy target for an opponent's weapon. How the soldier holds the knife and the mobility in his wrist dictate the type of strike he uses. The typical cut is an overhead chop with the edge facing downward at an approximate 60-degree angle. In combat it is particularly important to hold the kukri firmly and place full force behind the blow. As the knife moves through the overhead arc and toward the target additional power is acquired by relying on bodyweight and momentum, or by striking from a slightly elevated position if possible. Since the blade is swung in a downward motion, the force of gravity naturally increases the power of the cut. The free hand can be used to aid balance by stretching that arm slightly forward just prior to striking, while keeping the hand away from the knife's trajectory.

The distance when striking should be such that the arms can move freely, which equates to two to three feet from the target. To avoid accidental cuts in case of a missed swing, the arc through which the knife travels should end further away than its starting point or the knife should be wielded slightly upward on impact with the target.[98] Such cuts would prove exceedingly difficult to defend against, as one soldier explains:

> When we were engaged in the many wars in India, the Gurkha proved themselves our most formidable enemies . . . Brave as lions, active as monkeys, and fierce as tigers, the lithe wiry little men came leaping over the ground to attack moving so quickly, and keeping so far apart from each other, the musketry was no use against them. When they came near the soldiers, they suddenly crouched to the ground, dive *[sic]* under the bayonets, struck upwards at the men with their Khukuris or Kukris, ripping them open with a single blow, and then, after having done all the mischief in their power, darting off as rapidly as they had come. Until our men learned this mode of attack they were greatly discomfited by their little opponents, who got under their weapons, cutting or slashing with knives as sharp as razors, and often escaping unhurt from the midst of bayonets. They would also dash under the bellies of the officers' horses, rip them open with one blow of the Kukri, and aim another at the leg of the officer as he and his horse fell together.[99]

Although the kukri is primarily a striking/hacking weapon, it can be used in stabbing attacks (although, the broad tip and curved blade make it difficult to adhere to the line of thrust). Deep thrusting wounds will result in severe blood loss and loss of vital organ function. To build momentum for a powerful stab, the knife should be held in the rear hand with the cutting edge facing down. As the knife is thrust forward in the stabbing motion, the soldier provides additional power through a slight rotation and forward lean of his upper body, transferring his weight from his rear to front foot.[100]

As a combat weapon the kukri has seen its greatest use in jungle warfare. Thick vegetation, which tended to tangle longer close combat weapons such as swords and bayonets, could be cut with relative ease with the knife. Long enough to ensure tremendous cutting power and intimidation capability, but short enough to be wielded in tight spaces, the kukri also lent itself to the crowded environment of trench warfare as demonstrated primarily in World War I. During battle in Belgium the Gurkhas remained squatted on the ground, throwing "their kukris at the advancing enemy, and as soon as they hit them, pull[ed] back with the strings attached to them [the kukris]. Then they dash[ed] up fearlessly and account[ed] for many more."[101] When the Japanese entered India in World War II, the 37th Gurkha Brigade was ordered to break the Japanese resistance. In taking a hill the Gurkha platoons suffered light casualties, while the Japanese lost thirty-two men who were cut down in their trenches with the kukris.[102] The kukri also found significant use in the elimination of German sentries at night. For example, the Gurkhas serving in World War I carried no rifles after dark, but crept "up to the enemy lines in silence. The sentries would fall without a sound."[103] Before the Germans realized the intrusion, hundreds of them had been slaughtered. Wielding the kukri in daylight hours proved more difficult particularly against longer weapons such as swords that had a reach advantage.[104]

The Gurkhas received little formal training with the kukri. They were expected to know its basic handling characteristics through having grown up with the knife and through their familiarity with it as the national weapon of Nepal.[105] They had used the kukri in everyday tasks and learned how to handle it until it became a natural part of their being. They were also less likely than an untrained person to suffer inadvertent self-inflicted wounds. The kukri's unique design prevented one from wielding it as a sword or a typical knife such as a Bowie. It proved impractical for certain maneuvers such as quick handling, reversals of direction, slashes, or reverse grip tactics. The main training obstacle, however, lay not in teaching proper handling but in reshaping the soldier's mindset about the use of the knife. Rather than viewing it as a tool or aid to everyday living, he would begin to think of it as an effective combat weapon and learn to wield it with proper intent.

The rather crude training exercises included cutting bamboo trees of different sizes in order to strengthen and toughen the hands. (The cut bamboo was later used for building living quarters and latrines.)[106] Proper tightening of the muscles and joints resulted in the acquisition of power. In World War II, when soldiers had attained proper and clean cutting technique, they trained for hand-to-hand combat by jumping into trenches and running up hills. Kukri training also included partner drills against other weapons such as swords and bayonets, which the Gurkhas were likely to encounter in close range fighting.[107] Although the ideal was to parry the longer enemy weapon and land the killing blow while the opponent was struggling to recover his weapon, it did not always materialize. In the Burma Theater one Gurkha soldier was bayoneted three times in the shoulder, head, and leg before he reached for his pistol and fired six rounds into his Japanese counterpart.[108] Another Gurkha serving in North Africa in the battle of Akarit in 1943 slit the throat of an enemy sentry who probably never saw that attack coming, while his comrade dispatched four enemy machine-gunners, "two with his pistol, and two with his kukri."[109]

In modern warfare the Gurkhas consider the kukri as important as guns and grenades. Although the knife is a fearsome tool in the hands of a skilled soldier, it is more than a weapon of war. It symbolizes the respect felt for those who have fought for the honor of their regiments in the service of the British Crown. As described by a member of the Khukuri House, a large manufacturer and distributor of kukris in Nepal and a supplier to the British Gurkha units:

> For a soldier to have a kukri, or to wear it in the form of an emblem, he must achieve more than other soldiers by getting into and distinguishing himself in the Gurkhas. He must be deserving of that honor and prove

himself worthy of his kukri, from which in Nepal, he must never be parted, and he is under a moral obligation to use his kukri bravely in times of war. It is as symbolic to Gurkha men of war, and the Nepalese nation and its various tribes today, as are the spear to the Zulu and the sword to the Samurai. It would seem to stand for an idealized concept of bravery and method of fighting, for heroism and honor. It is a more honorable weapon, cleaner, nobler, braver and more dignified than the tank, because one must have the guts to engage one's opponent, eyeball-to-eyeball, in fair and equal combat.[110]

GURKHA RECRUITMENT PRACTICES AND MERCENARY STATUS

Despite their Nepalese origin the Gurkhas have served and continue to serve in the strictly western British forces. At the formation of the first Gurkha battalion in 1815, the lieutenant in charge of recruiting soldiers boasted that he had gone to the prisoner of war camps a single man, and come out of there a force 3,000-man strong. The battalion he raised from these recruits "became the 2nd King Edward's VII's Own Gurkha Rifles [which] serves the British still."[111] Most of the Gurkhas were from the eastern part of Nepal.[112] During the first four decades of recruiting, however, a mix of men were enlisted including Garhwalis and Kumaonis from the Indian Himalayas who were not strictly Gurkhas. This continued until some British officers suggested that only the Gurkhas came from a martial race, were sufficiently military-minded, and possessed the innate fabric of a good soldier. In 1883 the Government of India "ordered that Gurkhas be enlisted only in Gurkha regiments."[113] The result was that Gurkhas were often recruited from the same clans and families, and passed their legacy from generation to generation.

The recruit signs up for the purpose of serving "as an integral part of the British Army whilst retaining [his] Nepalese identity and culture, and adhering to the terms and conditions of Gurkha service."[114] The Gurkhas recruited in Nepal yearly were (and are) mostly young men from the hills, between seventeen and eighteen years of age.

Admission to the Brigade of Gurkhas and the selection process are fierce. The recruits go through a rigorous elimination process conducted by British and Gurkha officers. The problem lies not in finding enough men who are willing to fight, but in finding only the best in a group of thousands. One year from a pool 28,000 applicants, only 230 were chosen:

> Though the Royal Gurkha Rifles have dwindled in number, more and more Nepalese want to join up. Recruiters looking for as few as ten men have on occasion had more than a thousand show up, some as young as 14. (It is easier to fake your age as a Gurkha— the average height is 5 feet, 3 inches.) Many are lured not only by the mystique but by the pay, more than 12 times what they would make in Nepal. The training is rigorous and includes ten-mile hikes and running up mountainsides with over a hundred pounds of rocks on one's back. Needless to say, Gurkhas are famous for enduring long marches and can do so at different paces.[115]

Once in the British army, "they must serve for a minimum of 15 years, before they are discharged back to Nepal with a pension."[116] Landing a job as a soldier in the British army can lead to several positive life-changes. Not only is it a way out of poverty and a pension in old age, it is also a promise of adventure and an opportunity to see the world beyond the hill villages of Nepal. Some of those not chosen go on to apply with the Indian Army's Gurkha Brigade. Today India recruits approximately 2,000 Gurkhas per year (about ten times the number that Britain recruits) and maintains 46 Gurkha battalions totaling about 40,000 men, many of whom may fight against Muslim Pakistan.[117] One young Gurkha reports that his grandfather, father, and three brothers served the Indian regiments. It is really poverty that is the driving force behind their enlistment and competition is fierce in the Indian army as well, despite India's tradition of recruiting Gurkhas dating back to 1805.[118]

Prior to 1948 the Gurkhas were trained in India. Thereafter training has been conducted in Malaya (the Federation of Malaya was formed in 1948 and comprised eleven states including nine Malay states and two British settlements), Hong Kong, and after 1994 at the Gurkha Training Wing in Britain.[119] Although the initial British recruitment practices of Gurkhas included recruiting men who were fugitive criminals, the recruitment standards improved with time until only men from classes prescribed by the army headquarters could be recruited.[120] In World War II, however, the British were desperate and resorted to recruiting any loyal and able-bodied men. The 114,971 Gurkhas recruited from 168,294 applicants therefore escaped much of the rigorous screening. Thousands of inexperienced Gurkhas, Indians, Sikhs, and others joined the infantry divisions and fought in World War II jungle warfare.[121] Gurkhas today are employed both as peacekeepers and combat soldiers.[122] Although they retain their Nepalese citizenship, they are considered full members of the British forces and swear allegiance to the British Crown.[123]

A Gurkha trooper at Raffles City, Singapore during the 117th International Olympic Committee session in 2005. Note the kukri worn vertically along the small of the back. Image source: Huaiwei, reproduced under Wikimedia Commons license.

Why did (do) they do it? The Nepalese "hill men" grew up with little freedom of choice. They lived in the underprivileged sector of society, "until the opening of service under the British came the way of some and then only for the fittest." They had a chance to change life for the better—some desired simply to run away from home; others to earn a pension and return home at the completion of their service—although, for many it would be a change for the worse.[124] One Gurkha recruited in World War II admits that he never wanted to become a soldier but had plans to attend medical school in Delhi, India, "and maybe later study at Cambridge or Oxford in England. But the war shattered my dreams."[125] After World War II, ex-Gurkha recruiters called *Galla Walas* were sent to the villages to search for likely soldiers.[126] Some men joined the army as boy soldiers, as young as fifteen years of age. Others, illiterate and unmarried, felt that they had little else to look forward to and desired to leave the restricted life in the village. One man enlisted at the age of twenty-two, but when asked about his age lied and said that he was seventeen. Another Gurkha explains that he ran away against his family's wishes (many homes would go short-handed in the field if the young men left) and joined the army partly through peer pressure and partly because the recruiter had given him money as a bond.[127] A British officer who has served with the Gurkhas since 1985 confessed that in modern day, "[l]ower-ranked Gurkhas earn an average minimum salary of about 1,000 pounds a month, a fortune in their own country."[128] Others left their villages under great ceremony and well-wishes from family and friends, and were therefore ashamed of returning home a month later after having failed the tough entrance requirements. Some stayed in India or found other work in the Kathmandu valley.[129]

The lucky ones were enlightened by the experience, gained confidence and respect, and returned home richer both in money and spirit. The possibility of seeing the world beyond the mountains or of earning a *bahaduri* (bravery award), and the prestige associated with the award most certainly contributed to the desire to sign up for service in a foreign army. As explained by one driver in the Gurkha Transport Regiment in 1970, "Every man in this world wants to be brave and would like other people to call him brave. The Army is the easiest way by which one can demonstrate one's braveness to the world. It is this that makes men enlist in the Army."[130] But becoming a part of the British forces was not good enough by itself. One must also have an opportunity to confront an enemy on the field of battle. Since all conflicts did not mandate the use of military force—it might be years or even decades until an opportunity would present itself—the Gurkhas were particularly eager to engage an enemy when given the chance.[131]

A medal received for one's accomplishments in battle was valuable, because it was the only concrete proof of a man's bravery.[132] However, many years after the conclusion of the Malayan Emergency—during which many soldiers, although fortunate enough to escape with their lives, had to go on medical pensions due to severe injuries (some were discharged without a pension)—a former Gurkha explained that if he had to do it over, he would "much rather have no bravery award than an arm that had hurt him for the next fifty years."[133] Although some of those who had missed the opportunity to distinguish themselves in battle were annoyed when the war ended, most of the Gurkhas who had eagerly volunteered to serve Britain were just as eager to return to their homes and families.[134]

Controversy continues to circulate around the recruitment of Gurkhas into foreign armies, and some of the Nepalese political parties today are attempting to stop the recruitment of Nepalese men into both the Indian and British armies. According to an article published in *The Kathmandu Post*, notwithstanding international agreements (or the obsolete Tripartite Agreement), "any country could recruit its citizens into its Army to defend its sovereignty."[135] Although more than ten million Nepali speaking Indian citizens could serve in the Indian army, currently an estimated 100,000 Gurkha soldiers have chosen to do so. At the time of India's independence in 1947, the majority of Gurkhas chose to go to the new Indian army rather than serving in the regiments that were assigned to the British army.[136] (At the outset of World War II, the Gurkha component of the Indian army comprised ten regiments of two battalions each.[137]) An example of their recent service in the Indian army is the conflict in Kargil in 1999, which included Gurkhas in a fight against Pakistani infiltrators.[138]

Moreover, while the Gurkhas rose to become elite fighters, status gaps existing between the Gurkhas and those serving in the regular Indian and British regiments have sparked debate.[139] For example, all non-white troops serving in Burma in World War II lived in bamboo and canvas barracks, "while the British officers and NCOs [non-commissioned officers] lived in separate quarters made of brick and wood."[140] A conflicting view is that it is the norm for officers to enjoy better living quarters than regular soldiers, regardless of ethnicity or skin color. Regimental centers were established for the purpose of handling many of the administrative chores, for equipping the soldiers properly and providing basic training and physical fitness programs, and for providing comfort to returning soldiers.[141]

The mercenary status of the Gurkhas also brought problems concerning their understanding of the military mission. Not only did they experience exceedingly difficult combat conditions, including hunger and disease fighting the Japanese in jungle warfare in the India-Burma Theater in World War II, they often did not know the reasons for their fighting. Ignorance of the mission is more typical of mercenaries than of conscripts or citizen-soldiers. The Gurkhas were not fighting for an ideal, such as the preservation of western democracy. Rather, they fought for their battalion and peers, for their pensions, and often simply because their officers told them to fight. After having been taken to the recruiting depot in 1940, a Nepalese man claims to have had no idea what duty he was being enlisted for. After his head had been shaved, he hoped only that he would not be sent back home in disgrace for not meeting the requirements. (If the recruit passed a medical examination, he would be sent to a regimental center for training. If not, the *Galla Wala* would take him back home.)[142] Likewise, during the Malayan Emergency in 1951, a Gurkha claims to have had no understanding of why he was fighting the guerillas; he knew only that he was concerned with earning money and getting a *bahaduri* (bravery award).[143] Paradoxically the Gurkhas found fulfillment "in war against an enemy not theirs, in a foreign country, and under foreign officers."[144]

Serving in a foreign army under foreign commanders is no doubt a difficult undertaking. Mercenaries have historically enjoyed a mixed public image. Some have been feared, others loathed. Few have been respected.[145] The fame of the Gurkhas and the kukri rests in part with the idea that this non-western fighting force is one of the most impressive mercenary forces seen.[146] (The British Gurkhas and the French Foreign Legion may be the two most prominent modern mercenary regiments still in existence. Most countries nowadays have conscripted militaries or all-volunteer forces.) Under modern international law, however, "Gurkhas are not treated as mercenaries, but are fully integrated soldiers of the British Army operating in four units of the Brigade of Gurkhas, and abide by the rules and regulations under which all British soldiers serve."[147] Moreover, the Gurkha reputation for loyalty, which is not shared with other mercenary soldiers, has prevented them from "being labeled as mercenaries for hire."[148]

Proponents of the continued recruitment of Gurkhas into the Indian and British armies, argue that the Nepali Gurkhas are not considered mercenaries because, according to the Tripartite Agreement between Nepal, Britain, and India, they are not specifically recruited to fight an armed conflict but are employed due to a long-standing track record of service in the British forces. When Nepal closed its borders to the world between 1816 and 1951, Britain "gained near monopoly on the services of the Gurkhas."[149] Furthermore, they are not motivated mainly through private gain, but are treated "on the same footing as other units in the parent army." The agreement also stipulates that the Gurkhas should not be used in opposite camps in clashes with other Gurkhas.[150] The Gurkhas can thus be fully integrated both in the Indian and British armed forces.

In recent history, with Britain lacking real colonial interests, the Gurkha regiments have been scaled down to only a few thousand, with the last big drop in numbers occurring when Hong Kong was returned to China in 1997.[151] Today approximately 3,500 Gurkhas are serving Britain. Their time in training, on account of the recent urgency of the situation in the Middle East, has also been scaled down. For example, the jungle warfare training that normally takes nine months has been cut to only one month.[152] Debates continue around the disparity of pay and pensions; although, since 1997 the "take home pay of a Gurkha equals that of his British counterpart."[153]

Many of the British Gurkhas return home to retire, to the same villages that have provided thousands of fighters for the British Empire. It is likely, however, that soon these men who have served and died in modern wars, ranging from the trenches of World War I to the twenty-first century conflict in Iraq, will find no more work in the British army. The military remittances, which have allowed their villages to prosper, will then end.[154]

Members of a British Gurkha unit prepare to enter a building to clear it. Gurkhas of modern wars continue to carry the kukri as part of their kit. Note the soldier kneeling in the front row, his kukri carried vertically along the center of his back with the concave cutting edge to the left. If moved slightly to the left of the body, the knife can be drawn with a right hand cross-draw and be ready for immediate use with the cutting edge facing the adversary. Image source: Lance CPL C.D. Clark, United States Marine Corps, reproduced under Wikimedia Commons License.

"BRAVEST OF THE BRAVE," BORN OR MADE?

"The Commander-in-Chief of the Indian army at the end of the nineteenth century asserted that there is no comparison between the martial value of a regiment recruited amongst the Gurkhas or the warlike races of north India and one recruited among the effeminate races of the South."[155] Do some men inherently make better soldiers than others? Are some men culturally predisposed to soldiering? Do factors such as upbringing, consistent use of the kukri as a utility tool, and occupation of areas that saw constant conflict naturally "train" men to be better suited for warfare? Can high quality fighting characteristics be transported through one's genes to future generation warriors?

It has been suggested that the rugged Gurkha region of Nepal "bred" an exceptionally rough stock of people, and that fighting capacity depended both on race and hereditary instinct. Life in the hill villages was "drab, dull, and sometimes dangerous," as was war.[156] The harsh climate and terrain of the Nepalese hills tended to produce men who were sturdy in character and accustomed to enduring hardship. When equipped with their kukris the Gurkhas were traditionally skillful fighters, in part because they had grown up around the knife and consistently used it for every purpose "from manslaughter to peeling potatoes."[157] A man brought up under such harsh conditions is bound to be less likely to crack when far from base and outnumbered by the enemy, or when confronted by the elements of nature; when "cold, tired, wet, hungry, and afraid." In World War II hunger proved to be a constant problem made worse by captivity in German camps, where prisoners were given no rations for four or five days at a time.[158]

However, the Gurkhas did not come from a particular "warrior caste." Although sons followed fathers into soldiering for generations, they were not classified as soldiers but as "traders, cultivators, and herdsmen," who possessed an "innate sense of discipline" under stress.[159] Nevertheless, this quality contributed to their reputation for fierceness in war. A study of the general stress and psychiatric problems experienced by soldiers in United Nations infantry battalions of several countries, including Ghana, Ireland, Nigeria, Senegal, and Norway, reports that the Gurkhas had the lowest incidence of mental disorders compared to the Norwegians, who scored highest, it is believed, because of their respective differences in background and training. The Gurkhas were accustomed since childhood to certain socio-cultural and religious beliefs (for example, the belief in reincarnation and that dying in combat was a positive step of progression in the cycle of life), which battalions of other countries lacked.[160] In practicality, however, and in contrast to the Japanese or German soldiers' loyalty to their emperor or Hitler, the Gurkhas were driven to do their best not by a belief in a "divine" personality or supreme race, but by the *bahaduri*, the bravery award, the idea that one should not let oneself down or sabotage the good name of the regiment.[161]

No doubt were the Gurkhas men of good fighting quality, which was a reason why the British desired to include them in their forces. Some hold the view that the Gurkhas fought so well because of their "natural" loyalty (conditioned by their upbringing) to the British forces. Simultaneously, the British officers were thought to possess "natural" courage. A near perfect relationship was therefore created between the Nepali and British soldiers. A more plausible explanation for the Gurkhas' loyalty to the British might be the cultural boundaries that existed between them. The British had to acquire the trust of the Gurkhas before any successful action could take place on the field of battle. The Gurkhas fought well under British officers because these officers had made an effort to acquire the confidence of the Gurkhas by learning their language: Gurkhali. This effort fueled a sense of loyalty and respect because it helped convince the Gurkhas that "the British officers were genuinely interested in them."[162] It is also possible that the British were successful at attracting and keeping good fighters because of more pragmatic reasons such as the introduction of a bureaucratic welfare system into the Indian armies, and laws that governed discipline and punishment for any disloyalty among the soldiers. Men recruited from particular localities would also naturally form easier bonds with their comrades.[163]

Claims of loyalty and fighting quality cannot be a completely substantiated, however. For example, in 1866 a Gurkha in the Indian army was dismissed from service by a court-martial when he was found so drunk that he had failed to appear for inspection by the British officers. A reason why this type of offense did not happen more often may be because dismissal from the army was often considered the worst punishment imaginable, more severe even than lashes. The soldier depended on continued employment for the financial well-being of his family. Although desertion existed, it was due mostly to a need for the soldiers to go home to their villages for a few months each year to help with the harvest. The army solved the problem by granting the soldiers leave for three months out of the year. The relatively low desertion rate among Gurkhas might thus be related to their need for employment away from home rather than a high degree of loyalty.[164]

    Some say that the Gurkhas could stand any hardship, except abuse by their officers. Others claim that the Gurkhas were indoctrinated with "total commitment to duties, courage under fire, and unquestioned loyalty to Great Britain."[165] The Gurkhas themselves say that they appeared brave mainly because they had such bad tempers when something made them angry.[166] Although they were not without fear, they often found ways to deal with the fear. For example, when fighting the Germans in World War II, a Gurkha soldier intoxicated himself with grape wine until he "felt so full of vigour that the Germans seemed smaller than before." When going in for the hand-to-hand fight, he "cut down four or five men" with his kukri.[167] At other times bravery came as a result of a misunderstanding. During the Malayan Emergency, a Gurkha accidentally fell into a guerilla latrine. When he charged toward the river to wash himself off, the others, who had initially refused to obey the order to fight, thought that he was charging the enemy alone and felt obliged to follow.[168]

    The Gurkhas also had little to lose by risking their lives. They had joined the army under oath and believed that they had to be ready for whatever hardships had been destined them. One man put it this way: "If you think you'll get back home, you'll be no use in war."[169] Although some believe that the Gurkhas have the ability "to live in the present and make the best use of it,"[170] one Gurkha, when seeing an irrigation ditch like the one his family had at home, thought about his village and was overcome with homesickness until tears came to his eyes.[171]

Additionally, the noise, chaos, fatigue, and attrition of war will catch most soldiers unprepared. One Gurkha fighting in Java in the aftermath of World War II admitted that he eventually lost his fear of Javanese bullets, but not of bombs.[172] Another soldier acknowledged that his superiors had been wrong when they had assured him that Gurkhas did not suffer from shell shock. Explosions from Japanese artillery caused several Gurkhas to start "shaking, mumbling, crying and urinating and defecating in their pants."[173] After coming under heavy enemy fire during the Falkland campaign, one Gurkha reported feeling ill and found the experience no "fun at all."[174] Although the Gurkhas have a reputation for enduring hardship and even remaining cheerful in the most dire of circumstances, when removed from their natural element (the hills and jungle) and transported to uncharted territory (some got seasick during their transport to Europe), the fear could at times became difficult to control. One Gurkha later admitted that during parachute training he had been so scared that he had to be pushed out of the door, and some of his peers had wet their pants. Another soldier fighting in Burma nearly drowned during a swimming test.[175] Others had a reputation for being poor sailors, unable to tolerate rough seas. During their travel by ship from Britain to the Falkland Islands, "prostate

[Gurkha soldiers could] be found in cabins and other areas of the ship, as well as pools of vomit on every deck." Pills that countered sea sickness seemed to offer little help. One Gurkha Rifleman stated that he would willingly have "sat through hours of Argentine torture" rather than going "through another minute of this ordeal."[176]

Like all soldiers the Gurkhas had to succumb to the "fog of war" (uncertainty, chaos, and a general lack of situational awareness) and the elements of nature. In World War I along the Euphrates River the heat and humidity was crippling, "boats were pulled through swamps and the men fell in scores from heat exhaustion, dysentery, and malaria." One force was reduced from 800 to 350 men.[177] In the India-Burma Theater in World War II, where "more Gurkha battalions were engaged than in all other theaters combined,"[178] not only were the conditions nasty and equipment in short supply, monsoon season brought innumerable hardships including tropical diseases such as malaria, typhus, and dysentery, which incapacitated one-tenth of the fighting force. Tropical rain forests covered most of the northern half of Burma. Mountains, jungles, and rivers turned the area into an exceedingly complicated battleground. "Mosquitoes at night, leeches in the day, and the stench of rotting corpses" further challenged the fortitude of the young Nepalese soldiers.[179] Moreover, although the Gurkhas have been compared to terriers pursuing rats in their zest for the Japanese enemy, the Japanese resistance proved formidable.[180] The Japanese soldiers were well-equipped, trained in jungle warfare, disciplined and courageous, and prepared to fight to the last man. "They [the Japanese] had come—as they saw it—not as conquerors, but as liberators, with a divine mission to defeat the western colonial powers in Asia." Despite the belligerents' mutual bravery, fatigue was so severe that Japanese and Gurkha fell asleep in the same foxholes, half-buried in mud, too weak and sick to wield their weapons.[181]

Although the enemy, terrain, and climate in Europe differed from the Burma Theater, Gurkhas fighting in the western world experienced their share of loss and hardship. During their twenty-month long campaign in Italy, the 1/5th Royal Gurkha Rifles suffered "losses of more than a thousand men," which amounted to "more than its original complement."[182] The heavily fortified monastery at Monte Cassino provided a formidable obstacle to the Gurkhas' advance across the Italian peninsula, as one soldier notes:

> I particularly remember when the Gurkha Regiment (the bravest soldiers I ever met) launched an assault on the high hill at Monte Cassino in an attempt to capture the heavily defended German position at the monastery on the top of the hill. They were within about 20 metres of the top when they had to retreat. They had run out of ammunition and supplies and could not be got to them. Tears were in their eyes as they retreated through our line.[183]

It was hoped that the Gurkhas, experts in moving across mountain terrain, would succeed when they launched the main assault on the Italian monastery. Instead they suffered heavy casualties measuring in the hundreds. Another story tells how the Gurkhas refused to advance on one of the last bastions of German resistance in the city of Tavoleto, because of the heavily armed defenders around the bastion. When the British lieutenant in charge of the Gurkhas threatened to advance with or without their support, the Gurkhas, "[s]hamed by their behavior . . . swarmed after the officer brandishing bayonets, kukri knives and grenades." The fight that ensued was so intense that by dawn all Germans still alive had been taken prisoners and "only thirty Gurkhas were left unwounded." Combat was often brutal. During the assault on Passano Ridge in 1944, in terrain that was broken by gullies and forests and precluded conventional infantry tactics, the Gurkhas infiltrated German lines and overran the enemy's trenches while efficiently killing enemy combatants with bayonets and kukris.[184] When the battle came to close quarters, the Germans, who were of significantly taller builds, attempted to fend off the assault with their rifle butts, but to little avail against the kukri wielding Gurkhas.[185]

As the war progressed one Gurkha confided to his brother that he had witnessed too many deaths, lives "snuffed out by a bullet, grenade, mortar shell, bomb, bayonet, sword or kukri," and had lost his desire to kill.[186] Similar doubts about the desire to continue battle can be extracted from modern conflicts, for example, in Sierra Leone where the Gurkha Security Group (GSC), a company "made up primarily by ex-Gurkha fighters," was hired to guard corporate assets. In a rebel ambush in February 1995 they suffered heavy casualties and lost their local commander in the battle, who was later reported "eaten by the rebels and his body emasculated, as a warning to other would-be interveners. Forewarned, GSC broke its contract with the government and left Sierra Leone."[187]

A Gurkha of 4th Indian Division keeps watch on enemy positions in Alpi di Catenaia from high ground on Monte Castiglione, 29 July 1944. Image source: Loughlin (Sgt), No 2 Army Film & Photographic Unit, Imperial War Museum, reproduced under Wikimedia Commons license.

It has been said that courage is not the absence of fear, but the conquest of fear. According to one Gurkha soldier, "God made us Gurkhas to withstand hardship. Without the Gurkhas the British and the Indians would have been beaten."[188] In the end, however, the Gurkhas are only men; not invincible, but mortal like the rest of us.

CONCLUDING REMARKS

The Gurkhas, still carrying their "traditional 18-inch long curved knife, known as the kukri, into battle," have served as elite forces of the British for nearly two hundred years "in countless wars, campaigns, and battles, showing their outstanding bravery . . . and spirit to fight until death under the harshest of conditions . . . shivering with wet and cold . . . scorched by a pitiless and burning sun, uncomplaining, [enduring] hunger and thirst and wounds. . ."[189] The kukri is a symbol of the Gurkhas' legacy and lethality.[190] The Gurkha soldier "keeps his kukri as he keeps his honour—bright and keen."[191] Although the Gurkhas in Britain admit that in today's army their broad-bladed fearsome trademark knife "is used mainly for cooking," and modern Gurkhas use firearms identical to western technology, we tend to associate the Gurkha with the kukri.[192] The knife symbolizes the importance of the soldier as an individual: "Steel and spirit became one . . . our souls united with our kukris."[193] Was it primarily the weapon, then, or was it the warrior that contributed the most to the western world's fascination with this non-western fighting force? Did the Gurkhas become famous because of their "inherent" fighting characteristics, or did the kukri as a symbol of status and skill obscure and glorify the reality of combat?

Wild stories circulate around the Gurkhas. No doubt is the kukri in the hands of a Gurkha a fearsome combination. Finding enough factual evidence to satisfactorily support the stories is difficult, however. For example, the Indian Corps tends "to minimize the effectiveness of the weapon," and most information about the knife revolves around its usefulness for bush cutting.[194] During the entire Malayan Emergency (1948 to 1960), although many stories have been told, only one occasion has been properly recorded of a "Communist Terrorist" being killed with a kukri.[195] It is likely that the knife saw its greatest use in World War I trench warfare and against sentries at night, and that the British did not encourage its use as a tactical weapon. Furthermore, the downsizing of the British army in the post-Cold War years has resulted in less time for kukri training. The kukri had to give way to more modern weapons.

Still, the events of the past demonstrate that the British could raise indigenous troops and use them successfully as a projection of their power. The Gurkhas have consistently served in jungle, desert, and trench warfare on battlefields far removed from their native Nepal. A noteworthy contribution of the Gurkhas to world military history is their long service in foreign wars, ranging from China's Boxer Rebellion to United Nations' peacekeeping efforts in East Timor, Rwanda, and Lebanon, and "as bodyguards for the Sultan of Brunei, one of the richest men in the world."[196] The Gurkhas' "cheerful disposition even when wounded" contributed to the British troops' fascination with these warriors. Success is also attributed to good training and pride in their regiments. As early as the Indian Mutiny of 1857, Gurkhas refused to revolt and came to the aid of the British civilians.[197]

The Gurkhas have proven their versatility in combat; they have demonstrated that they can fight in many climates and terrains in all corners of the world, from Asia to the Falkland Islands and Afghanistan. They have served in the capacity of fighting "brushfire actions or full-scale war." In the 1960s, while still carrying their kukri knife, the Gurkhas began "training in tactical nuclear armaments."[198] The Gurkhas also portray a general acceptance of their fate. As demonstrated through the study of stress and psychiatric problems experienced by soldiers in United Nations infantry battalions mentioned earlier, the Gurkhas are less soft, less complaining than their western counterparts. An acceptance of their fate does not imply a belief that their destinies are preordained, however. Rather, it is the realization that they come from regions of poverty and do not have much to look forward to at home that drives them into military service. Enlisting with the British forces would at least give them a shot at a better life. According to Colonel John Philip Cross, who at the tail end of World War II was assigned as a leader of Gurkhas to the 1st Battalion of the 1st Gurkha Rifles based at the town of Dharamsala in northern India, Gurkhas differ because they are in the habit of exerting themselves; they are not afraid of work, and "successful military discipline is an amalgam of tamed and inflamed instincts."[199]

While the Gurkhas are internationally recognized for their military capabilities with the kukri and have been termed the "only soldiers who can win their battles on their reputation alone,"[200] and the knife certainly makes a contribution to their fearsome reputation, how they fight has little to do with genetics, or with color, caste or creed. Many Gurkhas fight magnificently; others fight less well.[201] Some are exceptional men who reached high rank without the benefit of having grown up in the rough hill villages of Nepal, and without having learned the useful lessons of self-reliance and toughness.[202] Although they go by the motto, "It is better to die than to be a coward," the Gurkhas would rather live than die. But as suggested by Colonel Cross, "You can't fight properly until you know that you are going to die anyway."[203]

Gurkha Memorial, Winchester Cathedral, Hampshire. Image source: David Spender, reproduced under Wikimedia Commons license.

This book has demonstrated that Gurkhas have received unparalleled attention and fame as fearsome warriors. As expressed by one Gurkha, however, in order that their reputation can survive for future generations, "it is not enough to let the name [of the Gurkha soldier] honour us, we must honour the name."[204] The Gurkhas' strength lay not in the actual use of the kukri, but in their mental composure and willingness to sacrifice for the British army.

NOTES

[1] See Mike Seear, *With the Gurkhas in the Falklands: A War Journal* (South Yorkshire, U.K.: Pen & Sword Books, 2003), 22.

[2] See Himalayan Imports, *Some History of the Khukuri*, Himalayan Imports, http://www.himalayan-imports.com/khukuri-history.html.

[3] See Lionel Caplan, "Bravest of the Brave: Representation of the Gurkha in British Military Writings," *Modern Asian Studies*, Vol. 25, No. 3 (Jul. 1991), 586.

[4] J. P. Cross and Buddhiman Gurung, *Gurkhas at War* (Mechanicsburg, PA: Stackpole Books, 2002), 15.

[5] See Laxmi Thapa, et al., "The Battle of Deothal and Bhakti Thapa," *Nepalnews.com*, Vol. 22, No. 24 (Dec. 27-Jan. 2, 2003), http://www.nepalnews.com/contents/englishweekly/spotlight/2002/dec/dec27/opinion.htm.

[6] Cross and Gurung, 287.

[7] Khukuri House, *The Honor of the Kukri*, Khukuri House, http://www.gurkhas-kukris.com/kukri_history/honor.php.

[8] See Cross and Gurung, 12.

[9] See also Gorkhas. According to an article published in the Kathmandu Post, the term "Gorkha" is used in the Indian army and "Gurkha" is used in the British army. "Gurkha" is said to be a mispronunciation by the British. See Rakesh Chhetri, "Gurkhas in Kargil and Kosovo," *The Kathmandu Post* (Jul. 19, 1999),

http://www.nepalnews.com.np/contents/englishdaily/
ktmpost/1999/Jul/Jul19/editorial.htm.

[10]See N. Therese Edwards, "The Gorkha Fighting Arts," *Black Belt Magazine*, http://www.blackbeltmag.com/archives/260.

[11]See Khukuri House, *The Gurkhas,* Khukuri House, http://www.gurkhas-kukris.com/gurkhas_history.

[12]See Caplan, 576.

[13]The History Network, *The Gurkhas* (audio recording, Apr. 7, 2007), http://cdn.libsyn.com/thehistorynetwork/206_The_Gurkhas.m4a.

[14]See Brigade of Gurkhas, *The Origin of the Kukri*, The Army Home Page, http://www.army.mod.uk/brigade_of_gurkhas/history/kukri_history.htm. The kukri also became popular with U.S. Special Forces serving around the world and has been used extensively, for example, in Afghanistan and Iraq. See Leroy Thompson, *Combat Knives* (London, UK: Greenhill Books, 2004), 47. It might be noted, however, that combat knives are used mainly for mundane tasks such as bush clearing and opening crates and cans. The knife has found less popularity in large-scale organized warfare where concealment is not the focus, and where long range weapons such as firearms have found greater utility.

[15]See Caplan, 571-573. Because of the long history of the Gurkhas serving the British Crown, sources available to western historians are primarily of British origin. When interpreting these sources, one must guard against potential bias.

[16]Kaushik Roy, "Military Synthesis in South Asia: Armies, Warfare, and Indian Society, c. 1740-

1849," *The Journal of Military History*, Vol. 69, No. 3 (Jul. 2005), 651 & 664.

[17]See Byron Farwell, *The Gurkhas* (New York, NY: W. W. Norton & Company, 1984), 28-29.

[18]The History Network.

[19]See Farwell, 28-29.

[20]See Khukuri House, *The Gurkhas*.

[21]See Maria Christensen, *The Legendary Gurkha Soldiers*, http://www.suite101.com/article.cfm/oriental_history/86225.

[22]See Caplan, 577-581.

[23]See General Board of Global Ministries, *Nepal*, http://gbgm-umc.org/country_profiles/countries/npl/History.stm.

[24]Caplan, 581.

[25]The History Network.

[26]See Christensen. The tradition of employing Gurkhas in the British army was further emphasized during the Indian Mutiny of 1857-1859, when Jung Bahadur, prime minister and virtual ruler of Nepal from 1847 to 1877, sent a contingent of Gurkha soldiers to aid the British. The Treaty of Sagauli was superseded in 1923 by the Treaty of Perpetual Peace and Friendship. This treaty confirmed Nepal's independent status until Britain's authority over India ended in 1947. In the late 1980s Britain began to phase out its employment of Gurkha soldiers, and announced in 1989 that it would cut the Brigade of Gurkhas by 50 percent. See Library of Congress Country Studies, *Nepal Relations With Britain*, http://memory.loc.gov/cgi-bin/query/r?frd/cstdy:@field(DOCID+np0124).

[27]See Himalayan Imports, *Some History of the Khukuri*.

[28]Farwell, 39.

[29]Ibid., 46-48.

[30]See Kaushik Roy, "Coercion Through Leniency: British Manipulation of the Courts-Martial System in the Post-Mutiny Indian Army, 1859-1913," *The Journal of Military History*, Vol. 65, No. 4 (Oct. 2001), 939 & 948.

[31]The History Network.

[32]See Manahadur Rai, "Gorkhali Ayo! Gurkha Soldiers in the Battle for Imphal, 1944," as told to Marty Kufus, *Command*, Issue 16 (May-Jun. 1992), http://stickgrappler.tripod.com/bando/c16.html.

[33]See Seear, 203.

[34]Neville Williams, *Chasing the Sun: Solar Adventures Around the World* (Gabriola Island, BC, CAN: New Society Publishers, Limited, 2005), 69.

[35]See Tim Ripley, *Bayonet Battle: Bayonet Warfare in the Twentieth Century* (London, UK: Sidgwick & Jackson, 1999), 45.

[36]See Time in Partnership with CNN, "War is Heaven," *Time Magazine* (Jun. 1, 1962).

[37]The Long, Long Trail, *The Battle of Neuve Chapelle, 10-13 March 1915*, http://www.1914-1918.net/bat9.htm.

[38]See Ripley, 41.

[39]See Farwell, 108.

[40]See The History Network.

[41]See Marty Kufus, "Make a Path for the Kukri," *Inside Kung-Fu* (Sep. 1993), 57.

[42]See Cross and Gurung, 164. "Ayo Gorkhali" literally means "Coming the Gurkhas," but is generally translated as "The Gurkhas are upon you." See Farwell, 13.

[43] Raymond Callahan, review of "Burma: The Turning Point," by Ian Lyall Grant, *The Journal of Military History*, Vol. 59, No. 1 (Jan. 1995), 169.

[44] See Kufus, 57.

[45] From World War One to present day, Gurkha regiments have won 26 Victoria Crosses. Individual Nepali Gurkha soldiers have won 13 Victoria Crosses while serving the British in France, Tunisia, Burma, Italy, Borneo, and Palestine. See Khukuri House, *Gurkha VC*, Khukuri House, http://www.gurkhas-kukris.com/gurkhas_history/ghurkas_vc.php.

[46] Brigade of Gurkhas, *The World Wars*, The Army Home Page, http://www.army.mod.uk/linked_files/gurkhas/The_World_Wars_and_the_subsequent_hisory.doc.

[47] See General Board of Global Ministries.

[48] Things Asian, *Union Jack Still Flies at Britain's Last Military Outpost in Asia*, http://www.thingsasian.com/stories-photos/3740.

[49] See Brigade of Gurkhas, *The World Wars*.

[50] See Brigade of Gurkhas, *Background Information*, The Army Home Page, http://www.army.mod.uk/brigade_of_gurkhas/history/brigade_background.htm.

[51] See Harsa Subba and Pradeep Meyangbo, "War Giving Families of British Gurkhas Sleepless Nights," *The Kathmandu Post* (Mar. 22, 2003), http://www.nepalnews.com.np/contents/englishdaily/ktmpost/2003/mar/mar22/index.htm.

[52] Allison Lampert, "Canadians Join Gurkhas in Search-and-Destroy Mission," *National Post, CanWest News Service* (Dec. 25, 2007).

[53] See Seear, 12.

[54]See Cross and Gurung, 287. During the Falkland War, the Argentines credited the Gurkhas with "magical powers." The Argentine Press described them as "a cross between dwarfs and mountain goats." See Nepalese Khukuri House, *The Brave Gurkhas*, Nepalesekhukuri.com, http://www.nepalesekhukuri.com/gurkhas.html.

[55]See Seear, xix-xxi.

[56]See Farwell, 289.

[57]See Seear, 270.

[58]Ibid., 73.

[59]Ibid., 225

[60]Ibid., 288.

[61]Ibid., 67.

[62]Ibid., 220.

[63]Ibid., 176.

[64]See Ripley, 41.

[65]See Victorino Matus, "Unleash the Gurkhas," *The Weekly Standard* (Oct. 2001), http://www.freerepublic.com/focus/f-news/545879/posts.

[66]See Cross and Gurung, 301.

[67]See Brigade of Gurkhas, *The Origin of the Kukri.*

[68]Himalayan Imports, *Some History of the Khukuri.*

[69]See Kufus, 57. The kopis had a wide forward curving blade which gave it superb chopping capability and made it a favored weapon by the cavalry. See Patrick Kelly, *Iron of the Empire: The History and Development of the Roman Gladius*, myArmoury.com, http://www.myarmoury.com/feature_ironempire.html. British explorer and soldier Sir Richard Francis

Burton describes the kopis as a sword of twenty-one inches in length and two and a half inches in width, with a "broad back and a wedge section," with the cutting edge "inside." See Richard F. Burton, *The Book of the Sword* (Mineola, NY: Dover Publications, 1987), 265. Note also the similarity between the kukri and the *yataghan*, or the Turkish sword of the sixteenth to nineteenth centuries, which was sharpened on the inner edge and used by infantry soldiers for cutting the throats of their enemies. See Richard Cohen, *By the Sword: A History of Gladiators, Musketeers, Samurai, Swashbucklers, and Olympic Champions* (New York, NY: Modern Library, 2002), 108.

[70] See Khukuri House, *Origin of the Kukri*, Khukuri House, http://www.gurkhas-kukris.com/kukri_history/khukri_origin.php.

[71] See Kufus, 58.

[72] See Farwell, 36 & 154.

[73] See Seear, 60 & 170.

[74] See Frederick P. Todd, "The Knife and Club in Trench Warfare, 1914-1918," *The Journal of the American Military History Foundation*, Vol. 2. No. 3 (Autumn 1938), 139.

[75] See Cross and Gurung, 62 & 72.

[76] See Rai, "Gorkhali Ayo! Gurkha Soldiers in the Battle for Imphal, 1944."

[77] A comparison of the shapes of edged weapons can provide insight into the primary use of the weapon. Of similar design as the kukri is the machete, which has a straight cutting edge and a half-moon shaped spine near the tip of the blade. The straight edge and the shape of the tip make the machete properly weighted for heavy chopping.

Sometimes kukris were carried in lieu of machetes. The kukri also became an accepted weapon for the Burma and Singapore Military Police. See Brigade of Gurkhas, *The Origin of the Kukri*. Another interesting comparison can be made between the boomerang and the kukri. Both weapons have similar design characteristics but dissimilar methods of deployment. Although some knives are designed as missiles in order to avoid engaging the enemy at close range, the kukri is not a throwing weapon. See Burton, 39.

[78]See Seear, 21.

[79]See Kufus, 58.

[80]See Nepalese Khukuri House, *The Kukri— The Blade of the Gurkhas*, Nepalesekhukuri.com, http://www.nepalesekhukuri.com/khukuri.html.

[81]Nepalese Khukuri House, *Khukuri Making*, Nepalesekhukuri.com, http://www.nepalesekhukuri.com/making.html.

[82]See Khukuri House, *The Honor of the Kukri*.

[83]Ibid.

[84]See Himalayan Imports, *Some History of the Khukuri*.

[85]Kufus, 57.

[86]Ibid., 58.

[87]See Edwards.

[88]See Himalayan Imports, *Training and Techniques*, Himalayan Imports, http://www.himalayan-imports.com/faq/Training.htm.

[89]Check-Six Online Museum, *Coin Made From Spent World War II Nepalese Gurkha Bullet Casing*, Check-Six.com, http://www.check-six.com/Museum/Weaponry-m.htm.

[90]See Cross and Gurung, 102. Many of the accounts involving the kukri's use in hand-to-hand

combat are unsubstantiated and rely on the memories of the individual soldiers.

[91] See Himalayan Imports, *Training and Techniques*. The best way to stop a determined attacker may be by severing the spinal column. A forceful blow to an enemy's head with the side of the blade or spine of the weapon, however, can cause sufficient trauma to end the fight. Severing a limb will result in heavy blood loss possibly followed by shock.

[92] See Bob Crew, *Gurkhas at War: The Terrifying True Story of the Most Deadly Force in the World* (London, UK: John Blake, 2004), 13.

[93] See Khukuri Knife, *Khukuri Making Process*, Khukuri Knife, http://www.khukuriknife.com/khukuri_making_process.htm.

[94] See Seear, 112-113.

[95] Ibid., 22.

[96] Kufus, 58. Sacred ceremonies also included forming a long column, each man thrusting his kukri into a small flame and sprinkling the holy ashes on his head. See Rai, "Gorkhali Ayo! Gurkha Soldiers in the Battle for Imphal, 1944."

[97] Kufus, 58.

[98] See Himalayan Imports, *Safety Considerations*, Himalayan Imports, http://www.himalayan-imports.com/faq/safety.

[99] From the Rev. Wood, *Travels in India and Nepal* (1896), see Nepali Exporter, *Khukuri: The Gurkha's National Weapon*, http://nepaliexporter.com/files/british_gurkha_knife.htm.

Kukris and Gurkhas

[100]See Khukuri House, *Khukuri Handling*, Khukuri House, http://www.gurkhas-kukris.com/khukuri.

[101]Great War, "At Work with Kukri and Lance," *T.P.'s Journal of Great Deeds of the Great War* (Nov. 1914), http://www.greatwardifferent.com/Great_War/British_Front/Indians_01.htm. It might be noted that the kukri was not designed as a throwing weapon; although, some have falsely suggested that the notch at the base of the blade was used as a sight for aiming prior to throwing the weapon. See Seear, 22. Moreover, a weapon thrown becomes useless to the wielder if it misses the target, particularly if he has no way of retrieving the weapon.

[102]See Rai, "Gorkhali Ayo! Gurkha Soldiers in the Battle for Imphal, 1944."

[103]Great War.

[104]See Kufus, 58.

[105]Ibid.

[106]See Rai, "Gorkhali Ayo! Gurkha Soldiers in the Battle for Imphal, 1944."

[107]See Kufus, 58.

[108]See Cross and Gurung, 51.

[109]Farwell, 188-189.

[110]Khukuri House, *The Honor of the Kukri*.

[111]Farwell, 33.

[112]See Seear, 120.

[113]Ibid., 76.

[114]BBC, "Gurkhas in Kent," *Inside Out* (Jan. 31, 2005), http://www.bbc.co.uk/insideout/southeast/series7/gurkhas.shtml.

[115] Matus. Starting in 2007, Nepalese women are gaining entry into the Gurkhas. The British Ministry of Defense recruitment laws must now comply with the United Kingdom's sex discrimination regulations. Since the policy change, approximately six hundred women, many of whom have worked as porters climbing the Himalayas, have received training in preparation for the tough entry exam. Upon passing the exam, which includes both academic and physical requirements, these women will carry their kukris and serve alongside of male Gurkhas, and can hope to make up to twenty times the average Nepalese wage. See Dan McDougall, "Women Set to Join the Gurkhas," *The Observer* (Jun. 24, 2007).

[116] The History Network.

[117] See Anthony C. LoBaido, "Britain's Himalayan Mercenaries: Teen-Age Gurkhas Strive for Appointments to Serve in UK Military," *World Net Daily* (Feb. 24, 2002), http://ads.wnd.com/news/article.asp?ARTICLE_ID=26578.

[118] See Randeep Singh Nandal, *Gurkhas Line Up to Join the Indian Army* (Aug. 6, 2006), Naxal Terror Watch, http://naxalwatch.blogspot.com/2006/08/gurkhas-line-up-to-join-indian-army.html. When King Gyanendra took complete control of Nepal in 2005, the recruitment to the Indian army stopped. Now, with Nepal on the road to democracy, it has restarted.

[119] See Khukuri House, *Recruitment and Training*, Khukuri House, http://www.gurkhas-kukris.com/gurkhas_history/recruit.php.

[120] See Caplan, 583.

[121]See Cross and Gurung, 31, and Rai, "Gorkhali Ayo! Gurkha Soldiers in the Battle for Imphal, 1944."

[122]See BBC, "Gurkhas in Kent."

[123]See Brigade of Gurkhas, *Background Information*.

[124]See Cross and Gurung, 16 & 18.

[125]Rai, "Gorkhali Ayo! Gurkha Soldiers in the Battle for Imphal, 1944."

[126]In modern day the British employ approximately seventy *Galla Walas* for the purpose of locating promising recruits from the Nepalese countryside. See LoBaido.

[127]See Cross and Gurung, 190, 211, 239 & 248.

[128]Things Asian.

[129]See Farwell, 82.

[130]Ibid. The value one placed on bravery (or *bahaduri*, bravery award) is reflected in the first names given to many Gurkha boys: Manbahadur, Danbahadur, Bhaktabahadur, Fatehbahadur, Birbahadur, Rewantbahadur, Ranbahadur, Dalbahadur, Gaganbahadur.

[131]See Seear, 36.

[132]Ibid., 128.

[133]Cross and Gurung, 211.

[134]See Farwell, 233-234.

[135]Rakesh Chhetri, "Gurkhas in Kargil and Kosovo."

[136]See Caplan, 572 & 593.

[137]See Cross and Gurung, 31.

[138]See Rakesh Chhetri, "Gurkhas in Kargil and Kosovo."

[139]See Caplan, 575.

[140] Rai, "Gorkhali Ayo! Gurkha Soldiers in the Battle for Imphal, 1944."

[141] See Farwell, 181-182.

[142] See Cross and Gurung, 21 & 61.

[143] Ibid., 183.

[144] Ibid., 18. Likewise, in World War I the Gurkhas had little idea of where they were going. One battalion reportedly "started sharpening their kukris when the train neared Calcutta." See Farwell, 87.

[145] See Mike Chappell, *The Gurkhas* (Oxford, UK: Osprey Publishing, 1993), 4.

[146] See Himalayan Imports, *Some History of the Khukuri*.

[147] The History Network.

[148] Nandal, *Gurkhas Line Up to Join the Indian Army* (Aug. 6, 2006), Naxal Terror Watch, http://naxalwatch.blogspot.com/2006/08/gurkhas-line-up-to-join-indian-army.html.

[149] LoBaido.

[150] See Rakesh Chhetri, "Gurkhas in Kargil and Kosovo." According to Byron Farwell, although the agreement has been honored, "Britain and India also agreed that they would not employ Gurkhas against other Hindus or against civilians [although predominantly Hindu, the Gurkhas also adhere to elements of Buddhism], and both countries have violated these undertakings." See Farwell, 262. For example, the Gurkhas fought in Java in the aftermath of World War II to help restore order. Java has a predominant Muslim population but is also the home to large numbers of Hindu.

[151] See The History Network.

[152]See Harsa Subba and Pradeep Meyangbo, "War Giving Families of British Gurkhas Sleepless Nights."

[153]Cross and Gurung, 34.

[154]See Williams, 69.

[155]Caplan, 583.

[156]Cross and Gurung, 27.

[157]Todd, 142.

[158]Cross and Gurung, 21 and 141.

[159]See Farwell, 21-22.

[160]See Seear, xxii.

[161]See Cross and Gurung, 27.

[162]See Kaushik Roy, review of "Imperial Warriors: Britain and the Gurkhas," by Tony Gould and "The Gurkhas: The Inside Story of the World's Most Feared Soldiers," by John Parker, *The Journal of Military History*, Vol. 64, No. 4. (Oct. 2000), 1206-1207.

[163]See Kaushik Roy, "Coercion Through Leniency: British Manipulation of the Courts-Martial System in the Post-Mutiny Indian Army, 1859-1913," 938.

[164]Ibid., 946 & 958.

[165]Rai, "Gorkhali Ayo! Gurkha Soldiers in the Battle for Imphal, 1944."

[166]See Farwell, 280.

[167]Cross and Gurung, 134.

[168]Ibid., 205.

[169]Ibid., 27.

[170]See Seear, 7.

[171]Cross and Gurung, 63.

[172]Ibid., 165.

[173]Manahadur Rai, "Fighting Nature, Insects, Disease and Japanese, The Chindit War in Burma," as

told to Marty Kufus, *Command*, Issue 26 (Jan.-Feb. 1994), http://www.pownetwork.org.

[174] See Seear, 264.

[175] Cross and Gurung, 64 & 122.

[176] See Seear, 123-125.

[177] See Farwell, 108.

[178] Ibid., 205.

[179] See Rai, "Gorkhali Ayo! Gurkha Soldiers in the Battle for Imphal, 1944."

[180] See Farwell, 223.

[181] Rai, "Fighting Nature, Insects, Disease and Japanese, The Chindit War in Burma."

[182] See Farwell, 199.

[183] BBC, "Wartime Reminiscences: A Soldier's Story," *WW2 People's War* (Nov. 9, 2003), http://www.bbc.co.uk/ww2peopleswar/stories/57/a2002357.shtml.

[184] See Ripley, 119.

[185] See Khukuri Palace, *Who Are the Gurkhas?* http://www.khukuripalace.com/Content/History/Gurkhas.php.

[186] See Rai, "Fighting Nature, Insects, Disease and Japanese, The Chindit War in Burma."

[187] P. W. Singer, *Corporate Warriors: The Rise of the Privatized Military Industry* (Ithaca, NY: Cornell University Press, 2003), 112.

[188] See Cross and Gurung, 122.

[189] The History Network.

[190] See Chuck Simmins, "The Gurkhas Have Arrived," *The Command Post* (Apr. 2, 2003), http://www.command-post.org/archives/003357.html.

[191] Brigade of Gurkhas, *The Origin of the Kukri*.

[192]See BBC, *Who Are the Gurkhas?* http://news.bbc.co.uk/1/hi/uk/2786991.stm.

[193]Rai, "Gorkhali Ayo! Gurkha Soldiers in the Battle for Imphal, 1944."

[194]Todd, 1424.

[195]See Cross and Gurung, 15.

[196]LoBaido.

[197]See Robert D. Kaplan, "Colonel Cross of the Gurkhas," *The Atlantic Monthly* (May, 2006), 82.

[198]Time in Partnership with CNN.

[199]Cross and Gurung, 16-17.

[200]Ibid., 287.

[201]Ibid., 17.

[202]See Farwell, 81-82.

[203]Kaplan, 84. The acceptance of the probability of death on the battlefield is a common theme in several cultures, and has resulted in the ability of the soldier to give his full body and soul to the purpose of the mission. This is one reason why the Japanese, whose samurai tradition embraced the *bushido* code and the loyalty to a lord or emperor, proved such tough opponents.

[204]See Seear, 82.

BIBLIOGRAPHY

BBC. "Gurkhas in Kent." *Inside Out* (Jan. 31, 2005). http://www.bbc.co.uk/insideout/southeast/series7/gurkhas.shtml.

........."Wartime Reminiscences: A Soldier's Story." *WW2 People's War* (Nov. 9, 2003). http://www.bbc.co.uk/ww2peopleswar/stories/57/a2002357.shtml.

Brigade of Gurkhas. *Background Information.* The Army Home Page. http://www.army.mod.uk/brigade_of_gurkhas/history/brigade_background.htm.

.........*The Origin of the Kukri.* The Army Home Page. http://www.army.mod.uk/brigade_of_gurkhas/history/kukri_history.htm.

.........*The World Wars.* The Army Home Page. http://www.army.mod.uk/linked_files/gurkhas/The_World_Wars_and_the_subsequent_hisory.doc.

Burton, Richard F. *The Book of the Sword.* Mineola, NY: Dover Publications, 1987.

Callahan, Raymond. Review of "Burma: The Turning Point," by Ian Lyall Grant. *The Journal of Military History*, Vol. 59, No. 1 (Jan. 1995).

Caplan, Lionel. "Bravest of the Brave: Representation of the Gurkha in British Military Writings." *Modern Asian Studies*, Vol. 25, No. 3 (Jul. 1991).

Chappell, Mike. *The Gurkhas*. Oxford, UK: Osprey Publishing, 1993.

Check-Six Online Museum. *Coin Made From Spent World War II Nepalese Gurkha Bullet Casing.* Check-Six.com. http://www.check-six.com/Museum/Weaponry-m.htm.

Chhetri, Rakesh. "Gurkhas in Kargil and Kosovo." *The Kathmandu Post* (Jul. 19, 1999). http://www.nepalnews.com.np/contents/englishdaily/ktmpost/1999/Jul/Jul19/editorial.htm.

Christensen, Maria. *The Legendary Gurkha Soldiers.* http://www.suite101.com/article.cfm/oriental_history/86225.

Cohen, Richard. *By the Sword: A History of Gladiators, Musketeers, Samurai, Swashbucklers, and Olympic Champions.* New York, NY: Modern Library, 2002.

Crew, Bob. *Gurkhas at War: The Terrifying True Story of the Most Deadly Force in the World*. London, UK: John Blake, 2004.

Cross, J. P. and Gurung, Buddhiman. *Gurkhas at War*. Mechanicsburg, PA: Stackpole Books, 2002.

Edwards, N. Therese. "The Gorkha Fighting Arts." *Black Belt Magazine.* http://www.blackbeltmag.com/archives/260.

Farwell, Byron. *The Gurkhas.* New York, NY: W. W. Norton & Company, 1984.

General Board of Global Ministries. *Nepal.* http://gbgm-umc.org/country_profiles/countries/npl/History.stm.

Great War. "At Work with Kukri and Lance." *T.P.'s Journal of Great Deeds of the Great War* (Nov., 1914). http://www.greatwardifferent.com/Great_War/British_Front/Indians_01.htm.

Himalayan Imports. *Safety Considerations.* Himalayan Imports. http://www.himalayan-imports.com/faq/safety.

.........*Some History of the Khukuri.*, Himalayan Imports. http://www.himalayan-imports.com/khukuri-history.html.

.........*Training and Techniques.* Himalayan Imports. http://www.himalayan-imports.com/faq/Training.htm.

Kaplan, Robert D. "Colonel Cross of the Gurkhas." *The Atlantic Monthly* (May 2006).

Khukuri House. *Gurkha VC.* Khukuri House. http://www.gurkhas-kukris.com/gurkhas_history/ghurkas_vc.php.

.........*Khukuri Handling*, Khukuri House. http://www.gurkhas-kukris.com/khukuri.

.........*Origin of the Kukri.* Khukuri House. http://www.gurkhas-kukris.com/kukri_history/khukri_origin.php.

.........*Recruitment and Training.* Khukuri House. http://www.gurkhas-kukris.com/gurkhas_history/recruit.php.

.........*The Gurkhas.* Khukuri House. http://www.gurkhas-kukris.com/gurkhas_history.

.........*The Honor of the Kukri.* Khukuri House. http://www.gurkhas-kukris.com/kukri_history/honor.php.

Khukuri Knife. *Khukuri Making Process.* Khukuri Knife. http://www.khukuriknife.com/khukuri_making_process.htm.

Khukuri Palace. *Who Are the Gurkhas?* http://www.khukuripalace.com/Content/History/Gurkhas.php.

Kufus, Marty. "Make a Path for the Kukri." *Inside Kung-Fu* (Sep. 1993).

Lampert, Allison. "Canadians Join Gurkhas in Search-and-Destroy Mission." *National Post, CanWest News Service* (Dec. 25, 2007).

Library of Congress Country Studies. *Nepal Relations With Britain.* http://memory.loc.gov/cgi-bin/query/r?frd/cstdy:@field(DOCID+np0124).

LoBaido, Anthony C. "Britain's Himalayan Mercenaries: Teen-Age Gurkhas Strive for Appointments to Serve in UK Military." *World Net Daily* (Feb. 24, 2002). http://ads.wnd.com/news/article.asp?ARTICLE_ID=26578.

Matus, Victorino. "Unleash the Gurkhas." *The Weekly Standard* (Oct. 2001). http://www.freerepublic.com/focus/f-news/545879/posts.

McDougall, Dan, "Women Set to Join the Gurkhas." *The Observer* (Jun. 24, 2007).

Nandal, Randeep Singh. *Gurkhas Line Up to Join the Indian Army* (Aug. 6, 2006). Naxal Terror Watch. http://naxalwatch.blogspot.com/2006/08/gurkhas-line-up-to-join-indian-army.html.

Nepalese Khukuri House. *Khukuri Making.* Nepalesekhukuri.com.
http://www.nepalesekhukuri.com/making.html.

………*The Brave Gurkha.*, Nepalesekhukuri.com. http://www.nepalesekhukuri.com/gurkhas.html.

………*The Kukri—The Blade of the Gurkhas.* Nepalesekhukuri.com.
http://www.nepalesekhukuri.com/khukuri.html.

Rai, Manahadur. "Fighting Nature, Insects, Disease and Japanese, The Chindit War in Burma." As told to Marty Kufus. *Command*, Issue 26 (Jan.-Feb. 1994). http://www.pownetwork.org.

.........."Gorkhali Ayo! Gurkha Soldiers in the Battle for Imphal, 1944." As told to Marty Kufus *Command.* Issue 16 (May-Jun. 1992). http://stickgrappler.tripod.com/bando/c16.html.

Ripley, Tim. *Bayonet Battle: Bayonet Warfare in the Twentieth Century.* London, UK: Sidgwick & Jackson, 1999.

Roy, Kaushik. "Coercion Through Leniency: British Manipulation of the Courts-Martial System in the Post-Mutiny Indian Army, 1859-1913." *The Journal of Military History*, Vol. 65, No. 4 (Oct. 2001).

.........."Military Synthesis in South Asia: Armies, Warfare, and Indian Society, c. 1740-1849." *The Journal of Military History*, Vol. 69, No. 3 (Jul. 2005).

.........Review of "Imperial Warriors: Britain and the Gurkhas," by Tony Gould and "The Gurkhas: The Inside Story of the World's Most Feared Soldiers," by John Parker. *The Journal of Military History*, Vol. 64, No. 4. (Oct. 2000).

Seear, Mike. *With the Gurkhas in the Falklands: A War Journal.* South Yorkshire, U.K.: Pen & Sword Books, 2003.

Simmins, Chuck. "The Gurkhas Have Arrived." *The Command Post* (Apr. 2, 2003). http://www.command-post.org/archives/003357.html.

Singer, P. W. *Corporate Warriors: The Rise of the Privatized Military Industry.* Ithaca, NY: Cornell University Press, 2003.

Subba, Harsa and Mevangbo, Pradeep. "War Giving Families of British Gurkhas Sleepless Nights." *The Kathmandu Post* (Mar. 22, 2003). http://www.nepalnews.com.np/contents/englishdaily/ktmpost/2003/mar/mar22/index.htm.

Thapa, Laxmi, et al. "The Battle of Deothal and Bhakti Thapa." *Nepalnews.com*, Vol. 22, No. 24 (Dec. 27-Jan. 2, 2003). http://www.nepalnews.com/contents/englishweekly/spotlight/2002/dec/dec27/opinion.htm.

The History Network. *The Gurkhas*. Audio recording. (Apr. 7, 2007). http://cdn.libsyn.com/thehistorynetwork/206_The_Gurkhas.m4a.

The Long, Long Trail. *The Battle of Neuve Chapelle, 10-13 March 1915.* http://www.1914-1918.net/bat9.htm.

Things Asian. *Union Jack Still Flies at Britain's Last Military Outpost in Asia.* http://www.thingsasian.com/stories-photos/3740.

Thompson, Leroy. *Combat Knives.* London, UK: Greenhill Books, 2004.

Time in Partnership with CNN. "War is Heaven." *Time Magazine* (Jun. 1, 1962).

Todd, Frederick P. "The Knife and Club in Trench Warfare, 1914-1918." *The Journal of the American Military History Foundation*, Vol. 2. No. 3 (Autumn 1938).

Williams, Neville. *Chasing the Sun: Solar Adventures Around the World.* Gabriola Island, BC, CAN: New Society Publishers, Limited, 2005.

Wood, Rev. *Travels in India and Nepal* (1896). Nepali Exporter. *Khukuri: The Gurkha's National Weapon.* http://nepaliexporter.com/files/british_gurkha_knife.htm.

About the Author:

Martina Sprague has a Master of Arts Degree in Military History from Norwich University in Vermont. She is the author of numerous books about military and general history. For more information, please visit her Web site: www.modernfighter.com.

Printed in Great Britain
by Amazon